Science, Democracy and Relativism

Science, Democracy and Relativism: The Production and Dissemination of Scientific Knowledge from the Viewpoint of Communitarian Epistemology

By

Haris Shekeris

CAMBRIDGE SCHOLARS

PUBLISHING

Science, Democracy and Relativism:
The Production and Dissemination of Scientific Knowledge from the Viewpoint of Communitarian Epistemology,
by Haris Shekeris

This book first published 2014

Cambridge Scholars Publishing

12 Back Chapman Street, Newcastle upon Tyne, NE6 2XX, UK

British Library Cataloguing in Publication Data
A catalogue record for this book is available from the British Library

ISBN (10): 1-4438-5337-2, ISBN (13): 978-1-4438-5337-8

The present work is dedicated to †Anthi and Sofia, Yiannis and Ioannis.

TABLE OF CONTENTS

List of Illustrations ix

Acknowledgements xi

Preface to the Reader xiii

Chapter One 1
Science and Democracy

Chapter Two 39
Lay Perceptions of Science and Scientists

Chapter Three 77
Science, Education and Citizenship

Chapter Four 119
Communitarian Epistemology and Science

Chapter Five 149
Meaning Finitism Defended

Chapter Six 193
An Example of a "Communitarian" Scientific Discipline

References 211

List of Illustrations

Illustration 1: "Image" of water molecules 89
Illustration 2: Frigg's schema of the elements of scientific modelling....... 97
Illustration 3: "Instantaneous" knowledge acquisition............................ 128
Illustration 4: Knowledge production in case of constative testimony 132
Illustration 5: Schema One: Causal links of knowledge production in a non-urgent scientific controversy 139
Illustration 6: Schema Two: Causal chain of knowledge production in the case of an urgent decision 141
Illustration 7: The iconic Pittas halloumi cheese package 163

ACKNOWLEDGEMENTS

I wish to thank all the persons that made it so that this work has reached the printing press. My Parents and Family, my Teachers and Friends, you all know who you are and I will be eternally grateful to you All. Thanks for Everything!

I would like to specifically thank my supervisor, Professor Alexander Bird, and my mentor during my later years at Bristol, Professor Emeritus Ken Binmore, for, dare I say, appearing (at least) to have shown some faith in me.

PREFACE TO THE READER

The present work is the result of the blending of many pre-sentiments rather than ideas that I have gathered since I began observing the world. In this brief introduction I will attempt to lay out some of these, in the hope that the reader will better understand the argument laid down.

A first guiding impression was the cacophony of languages encountered at the European School in Brussels, when one may learn to swear in most of the languages of the European Union before they reach puberty. At such a young age, the live encounters with children from different countries provided me with an appreciation of community diversity that has motivated me ever since and to which I sometimes think is akin to the methodological relativism employed by some anthropologists. Much later, this led me to a conversion to the idea of language relativity, which is roughly the idea that the structure of a language affects the way we conceive of the world, that is it affects our world-view. Hence the idea of communities that I advance here has linguistic communities as its progenitor. An advancement of this view may be of course to conceive of scientific communities as akin to linguistic communities and of their activities as forms of language games in a Wittgensteinian sense.

Puberty in a rather conservative, predominantly Christian Orthodox society such as urban Cyprus twisted my teenage rebellion into one against all forms of power and domination, rather than as one purely against God. In the dilemma between Jean-Paul Sartre and Albert Camus I chose to read the latter, whose portrayal of the absurdity of life somehow meshed with my local priest's message (at an age when my football-playing friends were reprimanding me for not identifying with Che Guevara and the Sunday-school kids were fed up with me asking questions all the time, I was lucky that my parish priest had the patience to listen to my questions and engage with me) of the imperfect nature of humans. This interplay between a good-natured religion and teenage rebellion helped me much later, when it gave me the power to question science's (or any institution's, for that matter, including religion) superiority in leading to the Truth in a predominantly positivist University of Bristol Department of Philosophy. Hence I became sceptical of the Truth and began supporting truths, and finally, in order to eradicate power-games (call it mistimed teenage romanticism), of truths constructed through consensus.

Unfortunately so far I have been unable to rid my life of power-games, indeed sometimes they seem to be the only spice of life. However, as many a postgraduate writing their magnum opus, I have found solace, as well as the will to continue my quest in communities such as the open-source software communities and projects like Ubuntu and Wikipedia. Such co-operative, well-structured communities have shown me that co-operation may be a genuine alternative. The mobilisation of students in Britain in 2013 showed me that this approach can be fruitfully carried through to the way politics is conducted, something which brings me to the final pre-sentiment, that of democracy.

As a teenager being fed with the past glories of Hellenism, I was given through my schooling a deeply distorted yet deeply utopian vision of democracy, drawing on the ancient Athenians for inspiration (ignoring for a moment other bits of our education, such as Thucidides, that portrayed Athenians as brute imperialists during the Peloponesian Wars). This, after a belated rebellion against monarchy during my first years in the UK, translated into the conviction that debate on societal issues, or in any case debate about issues that affect the well-being of the whole or a part of a society ought to be simple enough to be within the grasp of the well-informed lay-person, and that the lay-person ought to participate in the deliberation process as directly as possible. And that this should include issues that have a scientific element embedded. Otherwise one need not bother with democracy, instead preferring the rule of the experts, which, among other things, may prove to be more efficient and better in many respects. Somehow, in a manner which until today I fail to articulate clearly, this seems to me rather paternalistic and in any case not democratic, at least not democratic in the original sense of the word as rule by the people; hence the clash with the whole edifice of modern liberal democracy.

With these considerations in mind, I am now ready to offer this work to you, the reader. Keeping in line with the last paragraph, I hope that my argument is simple enough to be easily understood. Roughly, I argue that democracy dictates that no man's opinion matter more than anybody else's on a topic of societal concern – including experts (Chapter One). I then proceed to argue that current studies of perceptions of science favour a view of science favouring the formation of scientific elites, along with science communication efforts and popular science books (Chapter Two). Keeping in line with the democracy strand of the argument laid down in Chapter One, in Chapter Three I argue that education for citizenship and science education can reinforce each other in producing citizens well-enough informed to be able to adjudicate and exert control over technically-laden decisions. Furthermore, I begin to explore some of the philosophical implications of the epistemological view that I am proposing. In Chapter Four I venture into a lengthy defence of communitarian epistemology, the theory of knowledge that I am advocating, as one accurately describing both the production and the dissemination of scientific knowledge. Chapter Five consists of an original defence of a philosophical view on meaning associated with communitarian epistemology. Finally, Chapter Six marks a return into more practical territory, by sketching how a "communitarian" field of science would look like, through the example of sustainability science.

I hope the above is helpful enough to lead to a thought-provoking reading of the book.

Nicosia, July 2013

CHAPTER ONE

SCIENCE AND DEMOCRACY

1.1 Introduction

The topic of the present work is the relationship between science, democracy and epistemology. I offer a philosophical defence of communitarian epistemology as appropriate and conducive to a society which is truly democratic, and where scientists involved in policy are held in check from forming unelected and unaccountable elites.

In this chapter, by way of introduction, I will delineate the domain of my thesis and sketch the main motivations behind it. I will do this first by presenting a picture of science emerging from remarks by science popularisers and some standard rebuttals from the consensus among philosophers and sociologists of science. The topic is what science cannot be, thus highlighting the difficulties of providing a definition of the term "science". I will then show how science is indispensable to the modern state, by highlighting its role both as sustaining the economy and on policy-making. Following this I will move onto a discussion of democracy. Through the examination of different notions of democracy I aim to show the sheer contestedness of the whole domain, before moving back to examine the interface between science and democracy. I will look into different models of this interface, before hinting at my preference for a model in which the lay public is inextricably involved. I will lay down the foundations for an egalitarian model of democracy before finally elaborating on the role that scientists will play in such a model. This will lead me into a first consideration and linking together of communitarian epistemology, scientific knowledge and society and suggesting a look at lay perceptions of science and their origins, something which will be the topic of the second chapter.

1.2 Remarks by science popularisers and the consensus in philosophy of science

In this section I will contrast a picture of science emerging from assertions by practising scientists and popularisers with rebuttals from the fields of the sociology of science and the philosophy of science, thus showing how difficult it is to define "science".

1.2.1 Remarks by science popularisers

I will begin presenting a popular picture of science through two quotes from Lewis Wolpert's 1998 Medawar Lecture, in which he lays down a demarcation between science and technology. Wolpert asserts that "[...] science provides the best way of understanding the world in a reliable, logical, quantitative, testable and elegant manner. Science is at the core of our culture, almost the main mode of thought that characterizes our age." (Wolpert 2005, 1253)

Furthermore, in drawing the distinction with technology, he asserts that

> The distinction between science and technology, between knowledge and understanding on the one hand, and the application of that knowledge to making something, or using it in some practical way, is fundamental. (...) Technology is much older than anything one could regard as science and unaided by any science, technology gave rise to the crafts of early humans, like agriculture and metalworking. Science made virtually no contribution to technology until the nineteenth century (Basalla 1988). (Wolpert 2005, 1254)

Another remark comes from an introduction of a book celebrating science, entitled "The new optimists" (Richards 2010). In his introduction the editor, Keith Richards, notes that

> For professional scientists, living down the wilful distortions and extravagant promises made on their behalf by the popular press is an occupational necessity, but seeing science downgraded to just another belief system is harder to swallow. Scientists, like the rest of us, have plenty of beliefs, but the pursuit of science does not allow the luxury of indulging them at the expense of proper procedure.
>
> Science is a way of trying to understand and explain the way things work. It is driven by endless curiosity coupled with a determination not to be beguiled by easy answers." (Richards 2010, xiii)

A further attempt to philosophically ground scientific theorising and practice comes from Stephen Hawking's and Leonard Mlodinow's hugely popular “A briefer history of time” . Hawking and Mlodinow, conclude their Chapter that has to do with the nature of scientific theorising by identifying curiosity and the thirst for explanation as the underlying motive of science:

> But ever the since the dawn of civilization, people have not been content to see events as unconnected and inexplicable. We have craved an understanding of the underlying order in the world. Today we still yearn to know why we are here and where we came from. Humanity's deepest desire for knowledge is justification enough for our continuing quest. And our goal is nothing less than a complete description of the universe we live in. (Hawking and Mlodinow 2010, 18)

Before addressing the argument from curiosity, however, I will look into two more attempts at a philosophical grounding of science by scientists themselves.

The first is an attempt by Brian Greene, in his popular book “The fabric of the cosmos”. Greene states:

> And physicists such as myself are acutely aware that the reality we observe – matter evolving on the stage of space and time – may have little to do with the reality, if any, that's out there. Nevertheless, because observations are all we have, we take them seriously. We choose hard data and the framework of mathematics as our guides, not unrestrained imagination or unrelenting skepticism, and seek the simplest yet most wide-reaching theories capable of explaining and predicting the outcome of today's and future experiments. This severely restricts the theories we pursue. (Greene 2004, ix)

The last quote comes from Alan Sokal and his book-length polemic against what he terms “the postmodernists of the left, the fundamentalists of the right, or the muddle-headed of all political and apolitical stripes” (Sokal, back cover). In his foray into defining a coherent philosophical position vis-a-vis scientific knowledge and scientific methods, he states that

> For us, the scientific method is not radically different from the rational attitude in everyday life or in other domains of human knowledge. Historians, detectives and plumbers – indeed, all human beings – use the same basic methods of induction, deduction and assessment of evidence as do physicists or biochemists. Modern science tries to carry out these operations in a more careful and systematic way, by using controls and

> statistical tests, insisting on replication, and so forth. Moreover, scientific measurements are often much more precise than everyday observations; they allow us to discover hitherto unknown phenomena; and they often conflict with "common sense". But the conflict is at the level of conclusions, not the basic approach. (Sokal 2008, p.178)

I will now contrast these remarks with the broad consensus in the philosophy, sociology and history of science.

1.2.2 Rebuttals from the history and philosophy of science

1.2.2.1 The "scientific method" and its characteristics

In what follows I will expose a well-known argument against progress in science, as well as two short arguments rebutting some of the quite naïve, philosophically speaking, assertions concerning the manner of science. These two arguments reflect the current consensus in the philosophy and sociology of science.

A first common thread in the above quotation is the view of science as consisting of some sort of method, or "procedure". However, the idea of defining science in terms of a methodology, in terms of criteria that would delimit this "mode of thought", has been largely discredited for philosophical reasons. The quest for fixed methodological criteria unique to science has largely fizzled out in recent years, thanks to the critiques of Feyerabend (1993) and Laudan (1978). A more radical extension of this view (from outside of philosophy, this time) may be found in Jenkins (2007) who argues that science as a unified enterprise and the scientific method are essentially 19th century political constructs and that school science is a misrepresentation in that it ignores important philosophical, conceptual, and methodological differences between the basic scientific disciplines. In the philosophical literature, the naïve unity of science project which gained momentum with Oppenheim's and Putnam's paper "The Unity of Science as a Working Hypothesis" (1958) has conceded a lot of ground to weaker methodological or heuristic theses of unity ([Fodor 1974], [Wimsatt 1976]). On the other hand, the mid-nineties have seen the emergence of arguments for the disunity in science being grounded in the fragmented nature of the natural world itself, such as the patchwork picture of the scientific universe described by Cartwright (1999) or the pluralist picture described by John Dupré (1993). These pluralist approaches however still insist on their realist credentials, as opposed to other anti-realist, non-realist or relativist epistemologies and ontologies.

A second family of views reflected in the above quotations consists in describing science variously as reliable, logical, quantitative, testable and elegant. I will give a couple of arguments against these assertions, even though the sheer variety of practices adopted by paradigmatic natural sciences should be enough to deter knowledgeable people from making such generalisations.

I will focus on three arguments that counter two main claims referred to in the quotations above. The first argument that counters the assertion that the manner of science is reliable, in the sense that its results invariably drive us towards truth, is the sceptical argument of pessimistic meta-induction, which has been spelled out by Larry Laudan (1981). Laudan's argument roughly claims that we have no reason to believe that key terms of our best scientific theories genuinely refer, as the history of scientific ideas is full of entities which have been later dismissed as non-referring. Examples of this are entities such as the ether, caloric, humours, etc. As regards theories themselves, a similar argument may be run: the historical record is full of dead theories, so there's no good reason to claim that our current theories are on the right track as regards their claims of being genuine candidates for approaching the true fabric of the reality that surrounds us.

The second and third short arguments that I will now propose concern the assertions that science is somehow "elegant" and that this elegance proceeds in part from the harmony and logical character of their most cherished equations.

In order to answer the first part, one may derive a lot of material from physics, which is regarded by some as the most paradigmatic natural science. The laws of physics, as Cartwright (1983) points out, can be separated into phenomenological and explanatory ones, the first being more related with what really happens in experiments, whilst the latter are generalisations and abstractions designed for ideal conditions. As Cartwright claims, the former are usually the elegant ones which possess much explanatory power but apply only to idealised conditions and hence never explain experimental phenomena, whereas the latter are usually more "messy" and do not possess the elegant character of the former, but they have more success in assessing and predicting experimental data.

As concerns the logical part of science and the beauty derived from its explanatory laws and their "logic", one may countenance the claim that perhaps logic is not so much a given embedded in scientific procedures but is rather an empirical discipline which co-evolves along with paradigmatic successful theories. For example, Wilce (2006) states that "Some have argued that the empirical success of quantum mechanics calls for a revolution in logic itself", something which indicates that quantum mechanics may redefine how we see logic. Hence, a revision in some cherished laws of logic such as that of the excluded middle is required in order to be consistent with the results and theoretical baggage of quantum mechanics.

1.2.2.2 Aims of science

In his second quote, contrasting science with technology, Wolpert distinguishes between the aims of science and technology in the following way: science, according to Wolpert, has the aim of obtaining understanding and knowledge of how the world works, whereas the aim of technology is more practical and doesn't focus on systematic knowledge. The same train of thought is to be found in a more dramatic fashion in the Hawking quotation, whereas it is also a part of Richards' quotation. I will here argue that the two aims, the purely epistemic-curiosity driven one and the instrumental one, cannot be neatly separated, and that furthermore the claim that science is only concerned with knowledge is untenable. My argument will begin by the consideration that scientific inquiry has often been made possible by practical, or at least non-epistemic beliefs, needs and innovations. Through this I will show that considerations other than that of the quest for understanding and of knowledge of how the world works were necessary for science to even get off the ground. I will then argue that the position that science's sole aim is knowledge is untenable and also that there is no evidence that a quest for objective truth leads to progress, which seems to be the ultimate aim of the "knowledge as understanding of the world" view. Finally, I will present an argument that in any case the distinction between fundamental and applied research is relative to the given context. These considerations will serve to undermine the claim that a neat separation between the aims of science and technology can be made.

The first example of innovation being necessary for "science-as-understanding of the world" may be discerned by Cardwell's (1972) study of western science and technology. Whilst Cardwell locates the origins of science in the 17th century, (1972, 3), with the systematization and institutionalisation of scientific inquiry, he also hails the impact of two technological innovations (the weight-driven clock and the printing press) that form the "twin pillars of our modern civilisation" (1972, 12). He claims that the Newtonian notion of absolute mathematical time, and the dissemination of printed material that allowed the lively dialogue among scholars in fact enabled science to emerge. Another example of standardization as an important factor enabling scientific inquiry is given by Chang's (2004) study of thermometry, and the historical and other contingencies involved in finding the fundamental truth of the constancy of a boiling point for water, which in turn paved the way for thermometric scales. Finally, in a non-european context, Baber (1996) shows how the invention of the spinning wheel in England, which triggered the industrial revolution, as well as research into dyes in chemistry, were both a result of the British effort to displace India from its dominant position as the world's supplier of dyed textiles. Furthermore, Baber also gives an example of how in ancient India the development of mathematics and astronomy related intimately with the religious background and how mathematical problems were linked to practical problems such as constructing appropriate sacrificial sites. (Baber 1996, 27)

The relationship between "pure" scientific inquiry and technology may also stretch not only in the past, but also in the future. Kitcher (2001, 89) notes that

> we may look forward and recognize that there are readily envisageable ways of linking the results of inquiry (or the possible results if the inquiry develops in a particular foreseeable way) to practical projects that others could be expected to pursue.

A similar line of argument is often invoked in defences of projects such as the Large Hadron Collider, where defenders of the project being proposed often have to resort to past practical innovations that have been developed as a result of related spin-off technologies, in order to secure funding for research they would consider as "fundamental". In conclusion, the above examples show that the basis of the distinction that Wolpert makes, do not correspond either to the history of science nor to its current manifestations.

A further powerful set of arguments against the view that the aim of science is an overriding concern with knowledge and understanding of truths of the natural world comes from Kitcher's proposal of well-ordered science, whose aim is to uncover "significant truths" (Kitcher 2001, ch.7).

Kitcher explodes the view that the aim of scientific inquiry consists solely of context-independent epistemic considerations, by considering an analogy with cartography and showing that there is no such thing as an "ideal" map out of which all other maps emerge (2001, ch.5). Instead he develops the idea that science is concerned with uncovering significant truths, with significance being subject to pragmatic considerations, as well as contingent on decisions taken within science in the past.

Kitcher dedicates a whole chapter (2001, ch. 12) to the question whether the quest for truth as an aim of science ought to trump all other values. He links this idea to the Enlightenment secularization of the concepts of truth and rationality, which in turn build on older notions of truth as stemming from the Divine.[1] Kitcher explores different strategies that the defender of the objective value of knowledge-as-justified-true-belief could follow in order to convince their opponent, and finds them wanting. He objects to the argument from a natural disposition essential to humans towards knowledge by arguing that the concept of rationality that would have to be involved in such a defence would be so attenuated that it would scarcely lend any support to a method of systematic inquiry. Furthermore, an overwhelming portion of our everyday behaviour cannot be traced back to manifestations of rationality. He thus fails to find any good arguments for the claim that truth may lead to any progress in our values, concluding his exploration of this question with the claim that

> Behind the often evangelical rhetoric about the value of knowledge stands a serious theology, an unexamined faith that pursuing inquiry will be good for us, even when it transforms our schemes of values (2001, 166).

He thus urges us to dispense with this theological residue, claiming that "We need agnosticism all the way down".

Another way in which the debate about the relationship between science and technology has been framed is to be found in the distinction between "basic", or "fundamental" research on one hand, and "applied" research or science on the other. In practice, however, as Ziman (2000) shows, the complicated system of patronage which is currently in place makes the distinction meaningless, as one funding agency's basic research is another agency's applied research. Ziman further argues that whether some research is fundamental is always relative to some other research, and that being fundamental is not an objective property of scientific activity, but rather that it is a relational one (2000, 21).He states that "In appropriate circumstances, almost any research project might turn out to be 'fundamental' relative to some body of knowledge". (2000, 21)

I thus conclude this section by reiterating my conclusion that, contrary to the assertions of science popularisers, science cannot be separated from technology in terms of the aims of the two activities. Indeed, the term "technoscience" is being used by some authors to describe "ways of making knowledge that are also ways of making commodities, or such quasi-commodities as state- produced weapons." (Pickstone 2000, 113–114)

---

[1] See Campbell (1992, ch. 5 and 8) for a discussion of the link between truth and the Divine in early Christian and Medieval philosophy.

1.2.3 The difficulties of defining science

In the previous section I argued explicitly that neither the methods nor the aims of science are unique to it, and implicitly that the fact of diversity in the practices of science makes the task of defining science a very difficult one. A further difficulty in defining science is the issue of the institutionalization. Cardwell claims that the social institution of science only gains significance in the 17th century (1972, 3). However, this appears to be too rash a move with historical studies of the institutionalization of scientific disciplines showing that this process differed from country to country and from discipline to discipline; furthermore sociological, cultural and historical factors were involved and played a crucial role (Lenoir 1997). Again, it seems that the diversity of scientific practices makes it hard to talk of science as a single institution. Finally, it is worth considering the historical argument that the spread of science in Europe began to take place roughly at the end of the seventeenth century, and that widespread dissemination of scientific knowledge only began with compulsory mass education at the end of the nineteenth century. This was approximately two centuries after the Scientific Revolution. (Fuller 1997, 110)[2]

Finally, the social character of science must not be disregarded. Scientific knowledge often consists in claims in which there is consensus as regards their truth among scientists. Wolpert recognises this[3], stating that "Science, ultimately, is about consensus as to how the world works" (2005, 1257).

Given the above considerations and the sheer diversity of practices described as scientific, I will abandon for the time being the attempt to define science in terms of a conceptual analysis of what counts as scientific practice or as to what methods could be termed as scientific. Before moving on to examine the role of science in the 21st century, I will briefly entertain the thought that "science" may turn out to be a family resemblance concept, following Wittgenstein's use of the term. In this case, there is no characteristic common to all its instances, but rather uses

[2] The sheer diversity of methods and ideas of science may risk including the whole history of ideas, as Feyerabend remarks from a realist perspective: "A science interested in finding truth must therefore retain all the ideas of mankind for possible use, or to put it differently: the history of ideas is an essential part of scientific method" (Feyerabend 1999, 214).

[3] Even though it may be argued that the insistence on consensus may render his hard-nosed realism incoherent, since an eternal mind-independent truth as an aim of scientific inquiry is independent of any consensus.

of the word display “a complicated network of similarities, overlapping and criss-crossing” ([Wittgenstein 1953] cited in [Biletzki and Matar 2011]). The idea that science may be a family resemblance concept, enables us to unite the diversity of scientific practice, however realist philosophers may be a bit coy in accepting the associate notion of language games, as that leads into a relativism about meanings.

I now shift my attention to the role science plays in modern liberal, capitalist societies, in order to bring forth its significance for democracy, which is my main motivation for the current work.

1.3 Science in the 21st Century

In this section I will briefly highlight the role that science plays in modern societies. I will focus on two aspects of science, the role that it plays for the economy, and the role that it plays in solving societal problems. By doing this I hope to show that science and its products are a constitutive element of the social, economic and political life in the 21st century, and that it is inconceivable for us to imagine anything like the modern state without at the same time mentioning science and its products. Of course, it may be claimed that, on a broad understanding of science, it has always been part of the organisation of state administration, however I will elaborate on the 21st century and life as experienced now. The claim that science is indispensable in the economy and in state decision-making will serve to introduce the discussion of democracy, which will in turn introduce the idea of scientists involved in state decision making as members of a ruling elite in modern states.

1.3.1 Science and economic growth

It has been noted above that at least some researchers claim that “pure” science made no significant contribution to technology before the mid-nineteenth century. Despite my rejection of this claim (and of the accompanying distinction between pure and applied science), I have to accept that “pure” science played essentially no role in the Industrial Revolution, which dramatically changed the history of mankind.

Kealey (1996) notes that the major technological innovations in the mechanisation of the textile industry, as well as the successive improvements on the steam engine which was used both in the extraction of coal and its transportation through trains were carried out by

> uneducated, often isolated, men who applied practical common sense and

> intuition to address the mechanical problems that beset them, and whose solutions would yield obvious economic reward (1996, 68).

As further evidence, Kealey contrasts the fortunes of France and the United Kingdom in the 19th century. Whereas France had instituted science and was producing great theoreticians, it was the United Kingdom that rose to become the major economic power.

The major changes in the economy may have started in the household; it witnessed in the mid-19th century, according to Mokyr (Mokyr 2002, 151–157), not one but three Scientific Revolutions, which had as a result what may be labelled "the greatest demographic shock to Western demographic history (at least since the Black Death), namely the decline in infectious disease in the industrialised West after 1870 or so" (2002, 167).

Mokyr labels the three Revolutions as follows: the sanitarian and hygienic movement which began in 1815 and was correlated with the statistical revolution that led to the development of epidemiology in the 19th century, the germ theory of disease, which, although not new (Mokyr [2002, 184] claims that it was proposed in the sixteenth century), became widely accepted after the successes of Pasteur and Koch from 1865 onwards, and finally the knowledge that trace elements of certain substances were crucial to human health, and that these need to be supplied through diet, which became more widespread in the beginning of the 20th century after experimental work reporting the inducement of scurvy to guinea-pigs treated with a particular diet. These revolutions had as a result the dramatic reduction of mortality and the accompanying improvement in the quality of life of the population.

However, during that time, which was earlier identified as the heyday of academic science, a more significant revolution was taking place in the economics of scientific and technological change. In 1851, Queen Victoria opened the Great Exhibition of the Industry of all Nations, an event that signalled the rising importance of technological artefacts and commodities. By the 1870s the first corporate research and development laboratories were being set up in Germany, and by the inter-war period in the 20th century R & D laboratories had become institutionalised within firms, especially in the electrical and chemical industries (Coombs, Saviotti, and Walsh 1987, 9). It was also in the second half of the nineteenth century, that the organisation of industrial enterprises in the most industrialised countries began to change, with small family enterprises being increasingly replaced by larger firms which focussed on more than one activity and in which the division of labour played an important role. It is important to notice that technological change as embodied by the mechanisation of the processes of production and other aspects of trade such as refrigeration and transport technologies, was one of the factors that helped bring about organisational changes in the firm (Coombs, Saviotti, and Walsh 1987, 8).

Three final points are worth making on technological innovation as related to the economy. The first is that whereas earlier economists and economic historians regarded technological innovation as exogenous to the economic system, in that innovation was regarded as independent of economic factors, Coombs et al. (1987, 6) argue that common features of economic processes of innovation arise because technological innovation is regarded as endogenous to the socio-economic system. Secondly, and as a consequence of the first point, progressive division of labour and the systematisation of technological change result in an increasing instability and dynamism of the socio-economic system (Clark 1985, 39–40). In a way, technology breeds technology. This brings me to the final point, which is to point out the failure of the old Baconian "linear" model linking academic science to economic growth through new technology. Kealey (1996, 216) cites extensive evidence that in fact most technological innovations that make it to the market are improvements on older technology, and that only ten percent of such innovations are a direct result of basic science. Coombs et al. (1987, 4) concur with this view, claiming that studies have shown that there is "no direct correlation between R&D spending and national rates of economic growth". However, the importance of science, both basic and applied, is undisputed, and it is increasingly been seen as a source of solutions for societal problems. To this, the relation between state policy and science, I now turn my attention.

1.3.2 Science and the State

We live in a technoscientific world in which the products of science are the cornerstones of any society, from the provision of electricity and running water to the internet and linear accelerators for the treatment of cancer. As technoscience is so pervasive in daily life, it is perhaps unavoidable that it is increasingly being consulted for the solution of perceived social problems. Primary examples of this are the environmental crisis, the obesity crisis and the fight against various diseases, for which the solutions sought are invariably from the technoscientific domain. As technoscientific solutions are increasingly sought for by policy-makers, questions of public participation in, accountability and responsibility for social action are induced. As Nelkin remarks on these issues,

> Science policy has always involved such dilemmas, but the pervasive influence of science and technology – the visibility of their social impacts

> – has brought growing public concern about technological development and declining confidence in the bureaucracies responsible for technical decisions (Nelkin 1977, 11).

Furthermore, on the muddling of the boundaries between political and technical decisions, Nelkin remarks that

> the increased reliance upon technical expertise has helped to obscure responsibility for major social decisions, weakening the system of checks and balances. And policies, framed in technical language, are often difficult to comprehend or to control (1977, 12).

Before moving into a more detailed discussion of the role science should play in a democratic society, I will briefly address the question whether science can indeed solve social problems. In order to do this I will sketch two contrasting opinions on the subject, and trace the underlying notions of science that motivate these two opinions.

The first answer that I will sketch is the negative one. I will trace two routes towards this answer, the first exemplified by Weinberg's (1972) discussion of science and trans-science, the second by Sarewitz's (2004) discussion on what he labels "the excess of objectivity".

Weinberg claims that a whole class of questions which seem prima facie to be legitimate questions to ask of science, in that they are questions of fact and couched in scientific language, cannot find their answer from within science. Weinberg labels this sort of questions "trans-scientific" and gives as examples cases where science could provide in principle the right answer, given enough time and money for an experiment according to the experimental standards recognised as paradigmatic by science, however such a possibility would be blocked by considerations of practicality, or of very long time-scales, or even ethical considerations. Two other example classes of questions that Weinberg labels as trans-scientific are questions from the social sciences, where the objects of study are too complex and variable to be submitted to any sort of generalisations, and questions whose answers involve aesthetic and moral judgements. Two actual examples referred to by Weinberg, which I will modify slightly here, are the following; firstly, the effect of low dose radiation directly on humans, instead of experimentation using mice. The data that exists actually is derived from data gathered following the explosion of the two nuclear bombs in Nagasaki and Hiroshima, as ethical issues prevent direct experimentation using humans.

A second example comes from the calculation of very low risks but which, when actualised, cause major disturbances or catastrophes. Such events are the recent tsunami case that caused the destruction of the Fukushima nuclear plant, the eruption of the Eyjafjallajökull volcano in Iceland, which caused major disruption to airline travel and led to an interesting re-negotiation of allowed levels of dust within aeroplane engines. The answer to both questions posed transcends science and is firmly in the domain of trans-science.

Weinberg does claim, in a positive note, that although science cannot answer trans-scientific questions such as questions of risk, it can, however, limit their scope by minimising the risk through the improvement of the existing technologies and of current scientific knowledge. Hence, such questions would lose their pressing nature. However, other commentators in the field of science policy are less optimistic, some claiming that the most interesting questions in science are in fact trans-scientific (Barker and Peters 1993, 4).

A different route towards the same conclusion, namely that science cannot answer questions of policy, comes through Sarewitz's claim that science provides "an excess of objectivity". By this, Sarewitz defines the idea that the inability of science to prescribe human action is down to "not a lack of scientific knowledge so much as the contrary – a huge body of knowledge whose components can be legitimately assembled and interpreted in different ways to yield competing views of the 'problem' and of how society should respond" (Sarewitz 2004, 389).

Sarewitz claims that this phenomenon explains why, for example in debates about the environment, additional scientific evidence makes the debates harder, rather than easier to resolve. Sarewitz backs this claim up using the philosophical literature on the disunity of science, as mentioned above. Hence, his claim is that given the multiplicity of world-views within science, factual evidence may support any of a number of different positions on a decision prescribing human action.

However, the same move of linking the aims of science with contingent values, interests and choices of the people practising it can be used to promote the contrary claim that science can and does in fact prescribe social action (Kitcher 2001). A look at the historical record will show that both on a broad definition of science, and on a narrower (as the institutionalised activity that has become prominent in the last two centuries), technoscience has often been used to decide social, political and economic issues. Nonetheless, it ought to be recognised, as the historical record shows, that technoscience is by no means the only way of solving such disputes, and that furthermore it may not even be in a privileged position to prescribe good solutions. Some of the solutions given by technoscience are judged in retrospect to be good and useful, such as the use of radioactive materials in treating cancer, whilst others, such as the development and use of nuclear weapons, have in retrospect been shown to be less than ideal.

Given the increased reliance on science in policy decisions as a matter of fact, perhaps the question ought not to be whether technoscience can provide answers to policy questions, but the place it ought to have in the perceptions of citizens in a well-governed society, in order that it will contribute to overall prosperity and happiness. In order to steer myself into a position to address this issue, I first turn to a discussion of democracy itself.

1.4 Considerations on democracy

In this section, I will introduce what are widely accepted as the core principles of democracy, before briefly presenting and contrasting three models of democracy, those of aggregative, deliberative and radical democracy. This will serve to convey to the reader that the realisation of the core principles of democracy is by no means a trivial issue, but on the contrary it reveals significant philosophical differences in how democracy itself is to be perceived. I will then return to the topic of science, in order to build my case that scientists taking part in state decision making may be accused of being an elite in the heart of modern liberal democratic societies.

1.4.1 The core principles of democracy

I will follow Christiano (1996, 3) in delineating the ideals of democracy as popular sovereignty, political equality, and the right and duty to participation in open and fair discussion.

Popular sovereignty is the notion that the laws and policies that regulate the lives of people within a society are a product of a decision process that involves the whole of the citizenry in one body. Hence, every adult citizen has a say, as expressed both through their participation in the deliberative process and through their vote, in deciding the courses of action to be followed by the social whole. It is worth remarking that at this point I am making no assumptions about the mode of participation and I leave it open as to whether this participation is direct or indirect. Furthermore, popular sovereignty does not by itself entail majority voting, as it limits itself to stating that decisions ought to be taken by the people (or their representatives), with no mention of mechanisms.

The second ideal requires all citizens to be politically equal in the process of decision-making. Political equality has two dimensions, the first being the requirement that every citizen has a single vote and that issues are decided by the results of votes through simple majority. The second dimension of equality is that of control, or power over the decision-making process. Hence, every citizen ought to have the same degree of power over the legislative body, if such a body exists (Christiano 1996, 3). This condition, as I will discuss below, gets violated by elitist conceptions of democracy.

Finally, the third ideal of democracy states that every citizen has the right and obligation to participate in an open and fair discussion on the issues at hand (Christiano 1996, 3). This ideal is particularly pertinent in modern democratic societies, where the debates are often couched in technical language which is beyond the average citizen's powers of comprehension. The last point reflects a more general point about the ideals of democracy, the fact that they are something which societies ought to strive for if they desire a truly democratic system, rather than what actually happens in practice. However, a number of issues arise in the interpretation of these ideals, and the question of the role that they assign to the ordinary citizen. I will now move on to the exposition of three models of democracy which have been put forward by theorists in the 20th century, which conceive a significantly different role for the ordinary citizen, and rely on quite different philosophical assumptions. This will enable me to examine more closely the role of scientists in a democratic society in the following sections.

1.4.2 The "aggregative" model of democracy

The first model of democracy that I will look into is the "aggregative" model, which, according to one commentator, is "the understanding of democracy which has become dominant in the second half of the twentieth century". (Mouffe 2000, 81)

This model is said to originate in the work of Joseph Schumpeter, and more specifically in his seminal book *Capitalism, Socialism and Democracy* (Schumpeter 1976).

According to Schumpeter's conception of democracy, the election of leaders who will decide issues is prior to in importance to the deciding of political issues by the electorate. Schumpeter proposes the view "that the role of the people is to produce a national executive or government" (1976, 269) and defines the democratic method as "that institutional arrangement for arriving at political decisions in which individuals acquire the power to decide by means of a competitive struggle for the people's vote" (1976, 269).

The aggregative model as proposed by Schumpeter was developed by Anthony Downs, who further develops an economic theory of democracy. Drawing from market economics, Downs pushes the analogy between government and private enterprise, asserting that

> In effect, it (government) is an entrepreneur selling policies for votes instead of products for money. Furthermore, it must compete for votes with other parties, just as two or more oligopolists compete for sales in a market. Whether or not such a government maximizes local social welfare (assuming this process can be defined) depends upon how the competitive struggle for power influences its behavior. We cannot assume a priori that this behavior is socially optimal any more than we can assume a priori that a given firm produces the socially optimal output. (Downs 1957, 137)

The main features the aggregative model is claimed to involve are the following, as described by Saward (2000, 67): "(...) atomistic (liberal) individualism; secret voting on the basis of 'pre-given', 'non-deliberative' preferences; self-interested voting; and the absence of consideration of the 'common good'".

A key advantage and avowed aim of theorists promoting such a model was that they claimed to be providing a "purely descriptive approach to democracy, in opposition to the classical normative one" (Mouffe 2000, 82). A further advantage was that such a model acknowledges the pluralism of values and interests inherent in modern societies, thus being an advance of the classical accounts which conceptualised a more uniform society.

However, the above model and its accompanying conceptualisation of individuals as purely self-interested, as well as their diminished participation in the decision-making process, have attracted criticisms of elitism and lack of motivation for participation by ordinary people.

According to one commentator, within Schumpeter's model,

> Citizens participate in the process of competition by voting but since they know very little they are not effectively the ruling part of the society. (Christiano 2008)

This model stood more or less unchallenged until the end of the 1980s, when focus shifted into the notion of deliberation, which is the next model that I will now consider.

1.4.3 Deliberative democracy

The next model of democracy that I will briefly present is that of deliberative democracy. According to one of its main proponents, deliberative democracy, is roughly "an association whose affairs are governed by the public deliberation of its members". ([Cohen 1989] reprinted as [Cohen 2003, 342])

The deliberative model has been espoused by the two most important political philosophers in the Anglo-American and Continental traditions respectively, in the latter half of the 20th century. These are John Rawls and Jürgen Habermas. The extent to which their writings have influenced deliberative models of democracy leads one commentator, Chantal Mouffe, to talk about the Habermasian school, represented by the writings of Seyla Benhabib, and the Rawlsian school, represented in the writings of Joshua Cohen (Mouffe 2000, 84).

A key normative feature of deliberative democracy is democratic legitimacy, which Dryzek and Liszt describe in the following manner:

> For deliberative democrats, the essence of democratic legitimacy is the capacity of those affected by a collective decision to deliberate in the production of that decision. (Dryzek and List 2003, 2)

This is an advance over the previous model of aggregative democracy, which did not address questions of legitimacy. According to Knight and Johnson (1994, 278), aggregative models "lack the 'moral resources' required to generate and sustain legitimate collective solutions to politically contentious issues".

The key feature of this model, then, is that it shifts attention from the voting aspect characterising the earlier model into the deliberative aspect. It hence claims that through a process of deliberation in which all interested parties are present (either physically or through representatives), they may freely change their preferences given the presentation of the interests and preferences of the other participants. What is sought for is consensus, with majority voting taking place if consensus is not possible either because of irreconcilable differences or in a actual, practical situation because of time constraints.

A point on which Habermas and Rawls agree is that, at least for ideal situations (situations that Habermas terms as "ideal speech situations"), such a consensus is possible and may be arrived at. Indeed, Rawls defines his liberal principle of legitimacy in a way that reflects Habermas' view: "Our exercise of political power is proper and hence justifiable only when it is exercised in accordance with a constitution the essentials of which all citizens may reasonably be expected to endorse in the light of principles and ideals acceptable to them as reasonable and rational". (Rawls 2005, 217)

I will briefly expose one attack on deliberative democracy, which will serve as an introduction to what may be a more realistic model of the democratic process.

Mouffe (2000, 90-93), essentially charges that both Habermas and Rawls fail to adequately deal with the fact of value pluralism, which is a characteristic more easily discernible in modern multicultural societies. She contends that Habermas' first move, trying to delineate domains that should be excluded from democratic deliberation is doomed to failure and the conviction that universal principles can be found to ground what remains in the domain of politics is doomed to failure, as the domain of politics cannot be neutral as regards values. A second more forceful move, which consists in the formulation of a different species of rationality (such as for example what Habermas terms "communicative rationality") which is what comes to the fore in deliberative situations and that leads, in ideal situations, in rational consensus among deliberators, fails, according to Mouffe, because it does not take into account human practices. Taking a leaf from the late Wittgenstein, Mouffe asserts that

> The failure of current democratic theory to tackle the question of citizenship is the consequence of their operating with a conception of the subject which sees individuals as prior to society, bearers of natural rights, and either utility maximizing agents or rational subjects. In all cases they are abstracted from social and power relations, language, culture and the whole set of practices that make agency possible. What is precluded in these rationalistic approaches is the very question of what are the conditions of existence of the democratic subject. (Mouffe 2000, 95-96)

On the topic of citizenship, Mouffe claims that

> The view that I want to put forward is that it is not by providing arguments about the rationality embodied in liberal-democratic institutions that one can contribute to the creation of democratic citizens. Democratic individuals can only be made possible by multiplying the institutions, the discourses, the forms of life that foster identification with democratic values. (Mouffe 2000, 96)

Following on from this critique, Mouffe introduces radical democracy as a different model of democracy that builds on differences in forms of life, rather than attempt to trump over them as the other models seem to be doing. I now turn to the examination of this model.

1.4.4 Radical democracy

The last model that I will look into is the one of radical democracy, proposed by Laclau and Mouffe (1985), but with antecedents in the work of the political scientist Elmer Eric Schattschneider (1960). Schattschneider brings out nicely the inherent conflict in politics, by claiming in his preface that "The assumption made throughout is that the nature of political organization depends on the conflicts exploited in the political system, which ultimately is what politics is about." (1960, vii)

Mouffe, in proposing what she terms the "agonistic model" of democracy, recognises what she terms "the ineradicable character of the dimension of power and antagonism" generated by value pluralism. She claims that rather than being something external to preconceived identities, which can be eliminated or diminished in the context of deliberation, power is constitutive of social identities and social life. Hence, she claims, that

> (…) if we accept that relations of power are constitutive of the social, then the main question for democratic politics is not how to eliminate power but how to constitute forms of power more compatible with democratic values. (2000, 100)

Accordingly, democratic politics is not a set of practices aiming at the impossible act of consensus without exclusion (that is, consensus on universal principles) but rather it aims at "an us/them discrimination in a way that is compatible with pluralist democracy". (2000, 101)

Hence Mouffe defines liberal democratic tolerance, which she considers as one of the values of pluralist democracy, as treating those with different ideas as legitimate opponents, as "adversaries" with whom we share the common ground of "the ethico-political principles of liberal democracy: liberty and equality"(2000, 102). She goes on to say that even though there is this common ground, there is still disagreement as regards the meaning and implementation of these principles, and that this disagreement cannot be eliminated through rational means.

The aim of politics, from the viewpoint of the position that Mouffe labels as "agonistic pluralism" is to transform antagonism (conflict among enemies) to agonism (conflict among adversaries), not by eliminating passions from the political process but rather mobilising them and channelling them through a democratic design.

She claims that whatever conflict is produced through such procedures is qualitatively different to the aim of rational consensus laid down by deliberative democrats. Rather, what is produced through radical democracy is "conflictual consensus", which "exists as a temporary result of a provisional hegemony, as a stabilization of power, and that it always entails some form of exclusion." (2000, 103) As a result, closure and decision-making occurs by actively excluding other possibilities and, according to Mouffe, "one should not refuse to bear responsibility by invoking the commands of general rules or principles". (2000, 105)

I have now completed the exposition of different democratic models, and I hope to have conveyed a bit the sheer contestedness of the whole terrain of understanding democracy and the role that the citizen is called upon to play in different models, from passive observer and occasional voter in the aggregative model, to deliberator with a view towards rational consensus in the deliberative model, to a participant in the democratic "form of life" in the radical democracy model.

Holding on to the final remark by Mouffe about decision-making and responsibility in the absence of universal principles, I now move back to science in order to elucidate the connection between science and democracy that motivates this book.

1.5 Scientists and democracy

1.5.1 Elite rule and elitist conceptions of democracy

The question that I will start with concerns the role of experts in a well-governed society (or state), and more specifically to the role of experts in a democracy. Plato famously opts for the rule of experts, (putting forward an extensive framework in *The Republic*) dramatically putting the following words in the mouth of Socrates, and perhaps bitterly reflecting on the sentencing to death of his master by a jury composed by non-experts: "I shall be judged like a doctor brought before a jury of children with a cook as prosecutor" (Plato 2004, 128)

From a different viewpoint, and with his focus on "the good life", Aristotle also pretty much opts for expert rule in his preference for "aristocracy" in the sense of the rule of the best persons (Miller 2012). However, as Miller remarks, Aristotle's position is more ambivalent as he also considers favourably the rule of the many through the pooling of their wisdom. Indeed, Miller notes that

> The central claim is that the many may turn out to be better than the virtuous few when they come together, even though the many may be inferior when considered individually. For if each individual has a portion of virtue and practical wisdom, they may pool these assets and turn out to be better rulers than even a very wise individual. (Miller 2012)

Moving on to the beginnings of liberalism, John Stuart Mill comes across as another advocate of elitism or aristocracy, as, whilst defending universal suffrage, he wants the knowledgeable to have the right to cast more than one vote. Thus, Mill claims that

> If, with equal virtue, one is superior to the other in knowledge and intelligence or if, with equal intelligence, one excels the other in virtue the opinion, the judgement, of the higher moral or intellectual being is worth more than that of the inferior: and if the institutions of the country virtually assert that they are of the same value, they assert a thing which is not. (1993, 307)

As shown in the above quotation, Mill is quite explicit that not all opinions on a matter carry equal weight, with the opinions of learned people carrying more weight. This in turn implies that some people have privileged epistemic access to the common good. This assumption seems to beg the question as regards the need for democratic decision-making, since the determination of the common good and the policies that would be conducive to it are exactly what is at stake in democratic deliberation. Mill is faced with the following dilemma: if the common good is known and only accessible to a special class of people (the educated) then there is no need at all to consult the uneducated and hence an elitist (and in any case, not democratic) system is sought for; if, however, the common good is not given and nobody has special epistemic access to it, then everybody ought to have an equal say in determining it and in laying down the policies conducive to it; however this last option is incompatible with plural voting, which, as mentioned above is the suggestion that the "knowledgeable" be granted more than one vote when deliberating.

A final example of elitism posing as democracy is the bureaucratic aggregative model proposed by Schumpeter and examined above. For Schumpeter,

> collectives act almost exclusively by accepting leadership – this is the dominant mechanism of practically any collective action which is more than a reflex. (1976, 270)

Schumpeter is explicit that two main conditions for securing that the democratic method is a success are that there be a social class from which politicians be selected (1976, 291), and that there be a well-trained bureaucracy that will advise and in cases instruct the politicians (1976, 293). A further condition is to limit the scope of the application of the democratic method as much as possible (1976, 292). Hence it seems that Schumpeter's system does not in any way recommend the participation of the whole of the polity in the decision-making process. On the contrary it seems to want to limit the active parts in decision-making as much as possible.

1.5.2 Scientists as part of the ruling elite

The above-mentioned historical examples are illustrative of a general trend (even by theorists who are otherwise considered as democrats) for a tendency towards elitist rule, justified as being better for society or more efficient. Indeed, it seems that all four models referred to above want to reserve a privileged position for "the knowledgeable" within society. Given the prominent role of scientists in modern societies, and of the celebratory status which is reserved for scientific knowledge, I move on to consider whether scientists may be viewed as part of the ruling elite. Criticisms of scientists as an elite are sparse, with great social thinkers such as Karl Marx refraining from criticising the relationship between science and political power, whilst in the grip of the celebration of "pure science" as opposed to its adulterations and misconceptions (see ([Restivo 1988]).

However, the idea of scientists as an elite class, endowed with generous funds by the state and only minimally accountable to it has been a present from the time of the Scientific Revolution, as witnessed in the writings of Francis Bacon, especially in his unfinished fable "The New Atlantis" (1966). This elitist conception of scientists is reflected in Vannevar Bush's (1945) report on the place of science in a modern democracy, which, as noted above, ushered in a new era for the relations between state and science. According to Kitcher (2001, 139), Bush, in his insistence that disinterested research into fundamental questions by scientists who would be the best and perhaps the sole judges of their enterprise, did not move far enough from Bacon's elitist conception, despite his report being permeated at some places by democratic values (Kitcher, 2001, 142). Given that, as noted above, scientists are being increasingly consulted on matters of public policy and on societal problems, their role as regards the promotion or undermining of democracy is a pertinent question which deserves attention.

The claim that I will take as a starting point is that in modern, expertise-laden societies, the ideal of political equality is violated, because citizens experience a sense of alienation and loss of power over policy decisions, given the complex and technical nature of modern policy debates. Nelkin (1977, 12–13) thus observes that

> Indeed, participation as an ideology seems to be of growing importance just when technical complexity threatens to limit effective political choice. The declining influence of the citizen in an expertise-based society has thus become a pervasive public concern widely interpreted as a failure in the existing system of representation. It has called for a revitalization of democratic principles and their extension to areas of sectoral and national policy where they were not formerly applied.

Hence, scientists, as the agents capable of comprehending and taking part in technical debates, to the exclusion of lay citizens, are in the position to become the powerful bureaucratic class, who may on occasions dictate policy even to ministers seeking to produce state policy, as described by Schumpeter. In his seminal essay describing what he terms the "scientization of politics", Habermas concurs that the relationship between the politician and the expert has been reversed. He argues that in what he terms as the "technocratic model",which is the one associated with what has been termed above "the aggregative model" of democracy, "The latter (the politician) becomes the mere agent of a scientific intelligentsia, which, in concrete circumstances, elaborates the objective implications and requirements of available techniques and resources as well as of optimal strategies and rules of control." (1971, 63)

To be sure, Habermas supports a scientization of politics, however his support comes in the form of what he terms the "pragmatistic model", which, he claims has its origins in Dewey. Habermas argues that

> In contrast, the successful transposition of technical and strategic recommendations into practice is, according to the pragmatistic model, increasingly dependent on mediation by the public as a political institution. (1971, 68)

He then goes on to elaborate on the necessary communication needed to take place between scientific experts and the agencies of political decision, remarking introducing again the lay public as a necessary ingredient:

> The latter type of communication, however, can be institutionalized in the democratic form of public discussion among the citizen body. The relation of the *sciences* to *public opinion* is constitutive of the scientization of politics. (Habermas 1971, 69)

Hence the dyad of the ruling elite consisting of politicians and scientists need to be expanded to include the lay public, if we are to achieve a truly democratic science and a virtuously scientized politics.

Funtowicz and Strand bring forth the notion of mutual legitimation between the institutions of science and those of political power. In such models, which the authors claim are still prominent, governance is legitimated through its appeal to scientific rationality as a foundation, but also scientific rationality is legitimated through its role as the only legitimate practice which yields objective knowledge. The two authors make the same move as Habermas, that is breaking up this coupling of power with the introduction of the lay public through which the interaction is mediated.

As noted above, Habermas is a staunch believer in rationality and is convinced that rational consensus is achievable, or is an ideal worth striving for. Unfortunately I do not share that conviction, siding with Mouffe's model of agonistic democracy. Hence, before attempting to introduce an egalitarian model of democracy with an appropriate place for scientists, I will look into the literature in order to expose a typology of models of the policy and science interface, in order to identify elements of participatory democracy and radical democracy as laid down above.

1.5.2 Models of the policy and science interface

Funtowicz and Strand (2007) give a typology of a family of four conceptual models, plus a fifth one describing the science-policy interface as regards environmental issues. I will now briefly present these models.

The first model is what the authors label the "modern model", which, according to them, draws its inspiration from Enlightenment ideals; the model itself transmits its essentials to the three other members of this family. According to it, science determines policy by the production of objective, reliable and valid knowledge. The authors claim that mutual legitimation of the institutions of the state and of science is an idea central to this model. Governance is legitimated through its foundation on scientific rationality, but also scientific rationality is granted a privileged status as the sole provider of officially accepted knowledge. Values and subjectivity are the domain of politics, whereas factual issues are the sole domain of science.

This model however depends on some crucial assumptions that break down at least in the case of environmental issues: they assume that all science is "dry" science with no complexities, that uncertainties may be reduced to probabilistic risks, and furthermore that scientists themselves speak with a single disinterested voice and there are no disagreements between experts or conflicts of interest. Given that all three are present in environmental issues, this model can be easily seen to be inadequate at least for policy regarding such issues.

The second model regarding the policy/science interface is what Funtowicz and Strand label as the Precautionary Model, which can be considered as an attempt to preserve the modern model in an age of uncertainty. A principle of precaution is added to policy decisions, which both legitimises and protects the scientific input. There are many formulations of the precautionary principle, with a prominent one being the "double negative" one from the Rio Declaration on the Environment and Development (UNEP 1992):

> In order to protect the environment, the precautionary approach shall be widely applied by States according to their capabilities. Where there are threats of serious or irreversible damage, lack of full scientific certainty shall not be used as a reason for postponing cost-effective measures to prevent environmental degradation.

It shall be noted that this is a normative principle that is imposed onto scientists, and it is usually used when the scientific community is largely convinced that a potential harm or risk is real, however the statistical uncertainty over the perceived risk is just over the stipulated threshold of significance accepted by most scientific journals. The application of the precautionary principle however is important since it presents a significant break with the purely technocratic vision of the modern model.

The third model of the relationship between policy and science is the framing model. In this model, the framing of the problem and the "ownership" by a particular discipline become parts of the policy considerations that take place prior to the production of the scientific input into the policy decision. It may be viewed as an extension of the modern model in the face of the fragmentation of scientific knowledge into many academic disciplines. In a sense, different scientific disciplines become competing stakeholders which present their case for ownership of the problem to the lay stakeholders and the politically-motivated policy-makers. Funtowicz and Strand hail this first step towards the democratisation of scientific knowledge production and use for policy purposes, commenting that

> It can be seen as an attempt to acknowledge and somewhat redistribute the power balance between experts and lay-people: the non-scientific framing exercise that scientists often implicitly (and unselfconsciously) perform, is taken away from them and democratised, at least at a superficial macro level. (Funtowicz and Strand 2007)

However, such a model falls prey to accusations of vicious relativism, as it: a) has no resources to contest the accusation that the choice of frame is ultimately arbitrary or may be based on covert interests, which may result in the legitimation of the misuse of scientific knowledge and b) it has no resources to even identify that such a misuse has taken place, as every choice of framing is treated as arbitrary as the next. Where such a misuse takes place, it may result in the delegitimation and rejection of the scientific enterprise as a whole in the eyes of the lay public.

A response to this problem, which has been adopted by European Commission guidelines (EU 2002), has been to opt for pluralism, where possible. Hence, the Commission calls for a plurality of different disciplinary, institutional and social perspectives questioning as many underlying assumptions as possible based on different values. The assumption is that the opinion emerging from such a process will be robust in the sense that different biases may cancel each other out. Such a case of enlightened consultation may indeed result in intersubjective knowledge production. However, the framing problem remains; it is more a case of Plato's Philosopher King rather than a democratic process of decision-making, and it furthermore still assumes a privileged status for scientific knowledge over other sources of knowledge.

The final model based on the modern model and its Enlightenment ideals is the demarcation model, which may be said to the extension of the modern model in the face of the realisation that the settings of scientific knowledge production are no longer solely the property of the state, but rather they are often owned by and promote private, non-state interests. If it is acknowledged that knowledge production takes place within institutions who have their own agendas, then this context is sufficient to cast doubt on the shaping and the selection of data and conclusions reported. Although presented with claims in a neutral and objective language, the policy-maker may legitimately suspect that the experts providing advice are functioning as silent issue advocates.

In order to face this danger, a clear separation between the institutions and individual scientists who create the knowledge required and those who use it is sought for. The ultimate aim is, according to Funtowicz (2006) to make sure that scientists are protected from any political interference which may threaten their integrity, and that furthermore they are not made to inappropriately shoulder the political responsibility which should rest solely with the policy-makers.

An attempt at such an institutional demarcation has been the desire for a clean demarcation between risk assessment and risk management (Funtowicz and Strand 2007), which has stumbled at the point of separating inputs from facts and values. Another attempt is what

Spangenberg et al. (2012) term the "economic model" by which the discipline of economics has attempted the demarcation between science and policy through the attempted (since the beginning of the 20th century) distinction between normative and positive analysis. However, the latter distinction has also failed to yield policy-relevant economic analysis, instead veering off into discourses in the philosophy of science.

The failure of all four of the above models thus leads me to conclude that their espousal of the Enlightenment ideals of science is to be blamed, as it is a feature which characterises all four of these models. The fall from grace of this ideal, has, fortunately the consequence that it takes with it the aspirations of the scientific community to maintain an elite position within the policy-making process. Hence the final model which I will describe in the current work has equality, at least as regards scientists and policy-makers, as a structural feature. Furthermore, the unmediated connection between scientists and policy-makers, a connection which has "the experts' (desire for) truth speaking to the politicians' (need for) power" (Funtowicz and Strand 2007) is broken by the empowerment of the lay-person, whom environmental or health problems begin to affect in their everyday life, and who is seen as a necessary component of the policy-making process since it is them who are called to enact the policies seen as key to issues such as sustainable development.

The workings of such a process will demonstrate the key features of communitarian epistemology. Policy-relevant knowledge is not something that exists prior to the group coming together as a community, rather the opposite, to wit, the community decides which claims it will accept as "their" knowledge, and furthermore, once such a decision takes place then automatically all the stakeholders "own" that knowledge: they are both its co-creators and its co-owners. However, the coming into existence of a community consisting of experts from various walks of academia, from the humanities to the natural sciences, as well as lay participants and stakeholders, is by no means a trivial matter. I will explore this topic in the concluding chapter of the present work.

As elaborated in Sections 1.3.2 and 1.3.3 above, science and technology occupy a very important role in modern liberal democracies, both indirectly in the sense that the economy is largely based on them, as well as directly through the technical element of much of state decision-making. This pervasiveness of a practice that is quite opaque to the general public gives new urgency and at the same time a new angle on an age-old question on the topic of good governance. I will now lay down the theoretical bases for an egalitarian model of democracy, before exposing the ideal position that scientists would occupy within it.

1.5.3 An egalitarian model of democracy

I will propose an egalitarian model of democracy which gives citizens the privilege of selecting society's aims. This model will be built on the ideal of political equality, hence sidestepping the criticisms I have levelled against the elitist models described above. I will then return to the role of scientists, highlighting the role they ought to play in a democratic society, before introducing and briefly discussing the theme of communitarian epistemology in conjunction with democracy. Finally, I will address the question of the role of scientists and the perceptions of science by lay-people in a well-governed democracy, bearing in mind the ideal of political equality.

The model of democracy that I will build on in this section is based on Christiano's (1996, ch. 2) defence of democracy, and his accompanying normative conception of citizenship (1996, ch.5). Christiano's argument for an egalitarian defence of democracy proceeds in four stages, which I will briefly elaborate on.

Christiano begins his argument by founding democracy on the idea that each person's interests, as opposed to their judgements as to what is right and wrong, ought to be given equal consideration in choosing the laws and policies of a society (1996, 53). A main difference between the two is that interests are attributes of people, and hence not right or wrong, although he demands a minimum objectivity in the sense that a person can be wrong about their interests[4] (1996, 181). Christiano then claims that social institutions ought to be just in the sense that they ought to embody the principle of equal consideration of interests, which states that advancing the interests of one person is as important as advancing the interests of any other person (1996, 54).

Secondly, Christiano defines the subject matter of democratic deliberation as the choice of which collective properties are to be brought about. A property is defined as collective

> if and only if in order to change one person's welfare with regard to this property one must change all or almost all of the other members' welfare with regard to it (1996, 60).

---

[4] However, this does not imply that somebody else can be a better judge of a person's interests. Christiano believes that the agents themselves are still more likely to know about their own interests than others, even though discussion with others may elucidate these interests (1996, 182). This is to be opposed to Schumpeter's claim that what we are confronted with in the analysis of political processes is largely not a genuine but a manufactured will (1976, 263).

According to Christiano the characteristics of collective properties which make them highly interdependent – in the sense that what affects the interest of one agent, affects all of the agents – are that they are non-exclusive, in the sense that altering one person's life alters other people's lives too; they are public; they are inevitable in the sense that because of their contents they are necessarily configured in one way or another; finally they are alterable, in the sense that which of the alternative possible properties society is to obtain is the product of deliberation.

The third step in Christiano's argument is that interests can only be served through a procedure that is collectively binding. Given that collective properties are interdependent and there are conflicts of interest over them, there ought to be collective decision-making over them. In order to assure all citizens of their membership in an egalitarian and just society, Christiano claims that social institutions ought to be manifestly just to their members, asserting that standards of justice are in principle within the capacity of ascertainment of lay citizens. Hence only a just collective procedure would serve the purpose of binding all citizens.

Finally, Christiano claims that the principle of political equality requires that each citizen have roughly equal resources to affect the decision-making procedure, and this is cashed out in terms of every citizen having an equal vote and an equal obligation to have a say in determining the collective features of the society they live in.

1.5.3.1 Citizens as choosers of aims

Christiano gives a normative account of democratic citizenship that supplements his model of democracy. I will briefly outline this account and the demands that it makes of citizens, before proceeding to accommodate expert scientific opinion within this model. Christiano's normative theory of the role of citizenship (1996, ch.5) consists of three components, which I will sketch below.

The first component of the theory is the contribution citizens have the right to make to the political decision-making process. Hence, Christiano identifies three elements of decision-making, the ultimate aims of policy, the means to pursue those aims, and the compromises needed in case of disagreement in the decision of the first two. Each element is necessary and all of them jointly sufficient to describe political decision-making in modern democracies. The input of the citizen is primarily and solely in the first element of decision-making. The citizen has the right and duty of choosing the overall ultimate aims of policy, and to do this they ought to take a standpoint on the whole of society (Christiano 1996, 169), and furthermore bear the responsibility of choosing what is fair and unfair, as well as the general aims of policy in sectors such as education and foreign policy. However, citizens do not choose the means to achieve these aims, nor conduct the negotiations in order to reach the compromises needed. These, I will propose shortly, are the dominion of experts (scientific experts included) and politicians.

The second component of citizenship is the activities citizens ought to engage in, in order to fulfil their role as choosers of aims. These activities consist in putting pressure on the decision-making process through voting for political candidates and organising themselves in interest groups and unions, and in deliberating in order to inform themselves and each other of their interests and deepen their understanding and perhaps consensus on ideas of justice and the common good (Christiano, 1996, 178).

Finally, Christiano sets certain standards as regards the understanding of aims by the citizen and their democratic activities. As regards understanding, epistemic considerations of truth need to be weighted together with other social and political needs (1996, 187).

Considering democratic activities, standards are needed to ensure that articulate, reasonable, and discriminating (according to the different groups comprising society) conceptions of interests and justice are presented and debated (1996, 190).

1.5.3.2 Scientists in a democratic society

In this section I will describe the relation between the scientific community and the state in an egalitarian democracy of the type presented above. I will proceed by first identifying the four roles that scientists may play as regards policy-making, before opting for the one most suited to the model of democracy described above.

Before sketching the roles of scientists, it is worth noting that the relationship between science and public policy is bi-directional, in the sense that not only science is routinely an input into public policy, but also public policy is routinely an input into science policy. As Macgill (1993, 62–63) notes, public policy influences science,

> both in terms of influencing the multiple agendas of problems which attract scientific research interest and related resources and in terms of the inevitable boundary effects which influence the undertaking, perhaps even the outcome, of that research.

This bi-directionality plays an essential role in compelling scientists themselves to be opinionated in public policy. However, in what follows I will focus on the input of scientists into public policy and the roles that scientists may adopt.

I will lay down a classification of four roles that scientists may adopt as regards public policy adopted from Pielke Junior (2007). Pielke talks about the following idealised roles which scientists may adopt regarding policy decisions: the Pure Scientists, the Science Arbiter, the Issue Advocate and the Honest Broker. Pielke claims that all four of the idealised roles he describes are critically important and necessary in a functioning democracy (2007, 7), as long as scientists are conscious of their role. I will now briefly describe the four roles. First, the Pure scientist is the disinterested scientist who leaves policy decisions to political decision makers. According to Pielke, (2007, 15) this is more of a mythical stance, especially given the entanglement of science and policy described above.

The second role is that of the Science Arbiter. The Science Arbiter also steers away from providing answers to normative questions, focussing instead on factual questions that fall under the scope of scientific investigation. However Science Arbiters do engage directly with policy-makers by forming authoritative committees and organisations that provide expert judgement and council to decision-makers. This role seems consonant with the ideal role that the scientists would occupy in what Habermas (1971, 63) labels as the "technocratic" model, which has been rejected above as elitist.

The third role that the scientist may adopt as regards policy issues is that of the Issue Advocate. In this capacity, the scientist explicitly aligns themselves with a political agenda, and he focusses on the implications of their research for that particular agenda. The Issue Advocate is explicitly engaged with normative questions and seeks to openly participate in the process of decision-making. A species of Issue Advocate is the Stealth Issue Advocate, who invokes the historical authority of science and its objectivity in order to restrict the scope of policy choice, whilst at the same time claiming to be disinterested in politics (2007, 7). Issue Advocates may be used by policy-makers to legitimise policy decisions by granting them a scientific aura, in the framework of what has been described above as the mutual legitimation of scientific and state institutions. Hence, Liberatore (1993, 35) observes that

> In the US case, for example, concerning many and often vulnerable decisions by regulatory commissions, the main defence is to have plenty of scientific evidence on display. This is more authoritative in a political culture which respects "science" a lot more than it does civil servants.

The final role that Pielke sketches for the scientist is that of the Honest Broker of Policy Alternatives. The Honest Broker also engages actively in decision-making as opposed to shying away from it, but, as opposed to the Issue Advocate, they try to expand and clarify the scope of choice available to the policy-makers. Hence, the Honest Broker seeks to accurately map the landscape of possible courses of action, but leaves it open to the decision-maker to decide on the chosen course, by considering their own interests and values. It is this final role that I will defend as the role most scientists ought to play in an egalitarian democratic society.

Going back to the issue of democracy, Pielke aligns the role of the scientist as Honest Broker of Policy Alternatives with a model of democracy borrowed from political scientist E.E. Schattschneider. As noted above, such a model is reminiscent of radical democracy as described by Mouffe. In Schattschneider's model, as in Christiano's, the people do not know all the details of the policy alternatives themselves, however they survive and make their decisions by establishing "relations of confidence and responsibility" so as to "take advantage of what other people know" (Schattschneider 1960, 137).

Hence, revisiting Christiano's three components of decision-making, we can now identify lay-citizens, scientific experts and politicians with the roles of choosers of aims, choosers of possible and available means and seekers of compromises in case of disagreements. Scientists, as Honest Brokers of Policy Alternatives, function so as to provide politicians with the space of available solutions to given problems and to fully work out and clarify the implications of each alternative. These solutions by themselves do not compel decisions, since the decisions to be taken usually fall in the scope of trans-science, in the sense that in order to single out a solution both the science and the interests and values of the given society need to be taken into account. Having described a model of democracy that would be truly egalitarian and a role for scientists that would not bracket them off as an elite unaccountable to and unchecked by the lay public, I will briefly introduce the theme of communitarian epistemology and link it with democracy, before closing off the chapter by addressing the relationship between the lay public and scientists.

1.6 Communitarian epistemology, scientific knowledge and democracy

So far, I have attempted to show the point that most current models of democracy allow for scientists, or at least for those scientists that are implicated in state policy, to function as elites unaccountable to the lay public. I tried to combat this by developing a model of democracy based on equality, and by sketching a role for scientists that would exclude them from forming such an elite. In this section I will address the issue of elite rule versus an alternative derived from communitarian epistemology.

The key element here is contiguity with perceptions of how democratic decision-making takes place. Knowledge is and perhaps has always been regarded as empowering, and science is cherished as an institution of knowledge production. Furthermore, as argued above, the way we perceive the modern state inextricably involves the use of scientific experts in decision-making. My claim is that a perception of knowledge production in science that conceptualises Nature as giving away its secrets to those individuals or collaborators that are more daring with their experimentation, or in any case of a collective scientific rationality derived from tools such as mass collaborations and peer review cannot but lead squarely into the formation of elites. The reason for this, especially the case of peer review is the exclusion of non-scientists from the process of the creation and dissemination of scientific knowledge. This method and the scrutiny that scientists undergo in order to establish "truth" in their claims, or of scientists forming large collaborations to work on essentially non replicable experiments[5], is veiled in the belief that only scientists can evaluate each other's work. However, as regards decisions and knowledge-production exercises which are of utmost concern to society at large, it is imperative that citizens feel that they are co-producing knowledge pertaining to what are essentially also their problems. Furthermore, a division of labour that means that scientists produce objective knowledge and non-scientists take this work and use it to solve societal problems won't do, as this still results in elite rule by scientists or bureaucrats trained in bringing together the scientific and the political sense of such problems.

In order to combat the above notion, and anticipating the argument that I will advance in later chapters, two features of scientific knowledge production ought to be stressed: that it is first and foremost the property of communities rather than individuals, and secondly that the status of the possessor of knowledge designates a social status, much like "the possessor of money" does. The first claim serves to combat the notion of individualism in the sciences, and furthermore by mentioning communities it promotes the preferred organisation of such communities, which is that of egalitarian democracy within which competing world-views battle for domination. The second claim serves to demystify knowledge as well as

[5] For example, given that there are no other laboratories the same as CERN, replication is only indirect from the other accelerators which are not the built in the same way.

rationality, and reinforce its empowering message. However these two features of knowledge are what I will later identify as the two theses consisting the position of communitarian epistemology, as introduced by Kusch (2002). Going back to the production and dissemination of scientific knowledge, and the democratisation of the processes of science from the viewpoint of communitarian epistemology and egalitarian democracy, we are left with a view that has some similarities with radical democracy as proposed by Mouffe and elaborated above. The issue at stake is the status of scientific knowledge itself, as well as its possessors in the case where action is needed to be taken by the whole of society. A bold relativist view which will be implicitly defended in later chapters is the following: it is possible that a community of scientists knows P and defends it as true, whilst another subcommunity knows -P which is the negation of P. As regards the wider community, it may be claimed that they are largely agnostic of the truth of P or -P, or that in any case what counts for them as knowledge of P is not considered as "knowledge" of P by the community of experts. Hence, on topics of concern to the whole of society, that is to all the different subcommunities, egalitarian democracy dictates that there be a consensus as to whether P or -P is considered as knowledge and whether the whole of society possesses the same status of knowers of P or -P. In such a case, and in order to avoid any charge of elitism, it is important that a division of epistemic labour does not take place. The wider community should not espouse as "their knowledge" whatever scientists accept based purely on trust to the scientists, but genuinely accept the claims of scientists in the same way as scientists try to convince each other. Hence the view that peer-review is not sufficient to qualify as being democratic, but in a sense something along the lines of an open review including lay-people as reviewers should replace it, at least when the issues examined involve the wider society.

Given the present discussion of radical democracy and elitism on the part of scientists involved in public policy, the question of how this can be avoided presents itself naturally. In later chapters I will argue that this can be done by changing the way we perceive the production of scientific knowledge, and in consequence, a change into how scientific knowledge is produced. However, before moving on to this kind of questions, it is important to look into the actual perceptions of science as an institution and of scientists themselves by lay-people, something which I will turn my attention in the second chapter. I will briefly explain why lay perceptions of science and scientists are important before examining them in the next chapter.

1.7 Conclusion: Public perceptions of science and of scientists

Given the important role that science plays in areas such as policy and the economy, and given the above egalitarian model which proposes for the public at least the role of choosers of aims, it is important that the public recognise in the discourse of scientists when they are functioning more as Issue Advocates (or, worse, Stealth Issue Advocates) or when they are functioning as Honest Brokers of Policy Alternatives. Furthermore, as most policy issues are accompanied by both a lack of consensus on the underlying values and interests at play and a high uncertainty as regards the scientific knowledge involved, according to Pielke efforts by scientists to adopt the Science Arbiter or Pure Scientist role often result in Stealth Issue Advocacy (2007, 94). In order not to fall prey to Stealth Issue Advocacy, which seeks to restrict the decision-makers' choice of course of action by labelling only a small number (preferably one) alternative as the only "scientific" alternative, and presenting it as stripped of values and interests, the public ought to have a sound perception of science as essentially bound with values and interests and enmeshed rather than divorced from everyday life and as of the same kind as democratic argumentation and deliberation.

Furthermore the lay public ought to be aware of and be able to distinguish between scientific and trans-scientific questions, so as not to force scientists to give answers to problems they can't address. In order for this to happen the public ought to have a clear grasp of the contested nature of scientific knowledge as well as of its social nature, and an epistemological model that will be conducive to a healthy, critical and democratic citizenship.

My task thus in the next chapter will be to chart existing perceptions of science and scientists, in order to pave the way for an epistemological model that will fulfil the above conditions. I will first look into ways of gauging perceptions of science by lay-people through large-scale questionnaire surveys, before criticising the whole enterprise and exposing how a romanticised picture of science is promoted through popular science books. In the next chapter I will begin my proposal for an epistemology that both does justice to scientific knowledge production and dissemination and to democratic aspirations, by looking at the role that science education and education for citizenship may play in forging democratic citizens. Having introduced communitarian epistemology, I will look into some of its philosophical implications through a case-study on the philosophical and educational aspects of scientific modelling.

Having introduced communitarian epistemology, in chapter four I will describe how it describes scientific knowledge production and dissemination, as well as how it gives a good account of how policy decisions with a technoscientific element are made. In chapter five I will defend two components of communitarian epistemology, those of meaning finitism and of the community thesis on normativity. Finally, in chapter six, I will return to the topics of democracy, science and epistemology, by describing sustainability science as an example of how science/policy may be co-produced by both lay-citizens, scientists and politicians.

CHAPTER TWO

LAY PERCEPTIONS OF SCIENCE AND SCIENTISTS

2.1 Introduction

In the previous chapter I highlighted the problem that unelected elites, be they bureaucratic or knowledge elites, pose to an egalitarian democratic system. More specifically, I flagged the danger of the scientific community becoming such an elite, and concluded that in order to avoid such an event, it is important that lay citizens have a good understanding as well as accurate perceptions and knowledge of what science is and what it is that scientists actually do.

In this chapter I will engage with literature on the subject of lay-people's understanding of the processes of science and of their attitudes towards science as an institution and of the people who conduct it. There are a couple of questions that guide, at least implicitly, the first part of this chapter. These are the following: a) is there, in practice, transparency over how scientific knowledge is produced, at least when it comes to solving societal problems? and b) is the trust that lay-people show towards scientists and science as an institution justified, in the sense of having a good grounding on knowledge of the processes associated with knowledge production in scientific settings?

The topic of transparency addresses the question of what exactly the relationship is between the subgroup of scientific experts on a given societal issue and lay-people, if the decision to be taken is to be seen as being taken collectively in a democratic manner. There are two ways in which this relationship may be construed.

The first is through the division of intellectual labour in a Durkheimian fashion. This consists in the lay public delegating the responsibility of knowledge production to the expert-scientists and acting on their recommendations. It is also claimed (Bird 2010, 39) that the delegation of intellectual labour is sufficient for knowledge produced in scientific fashion to count as social knowledge, that is as knowledge of the whole of society, due to reasons of division of intellectual labour and of organic solidarity between the knowledge-producing group and the lay-people.

This is the usual route actually taken, however, there is a problem of transparency associated with it: together with the delegation of the responsibility of knowledge-production comes also the delegation of scrutiny on the specific topic and associated recommendation. Only other experts have the know-how to question the scientific recommendation; the public simply lacks the mental resources to be able to pose genuinely probing questions on the terms of the scientific experts. Knowledge is widely available in university libraries and scientific journals, hence, according to this view, it is in principle there for the lay-person. Furthermore, the lay-person is in sufficient communion with such knowledge when they use its products, such as technological appliances (Bird 2010, 47–48). Such a formula can work if there is a high degree of trust on the part of lay-people towards scientists, however this trust ought to be grounded on a sound understanding of the processes and limitations of scientific knowledge-production, something which I will probe through the surveys that I will look into in the following sections. Even so, there remains a worry that such a procedure as described above is not really democratic, or in any case not a participatory species of democracy as the one described in Chapter One.

The second way in which the relationship between the scientific expert subgroup and lay-people may be construed is through the dissolution of the epistemic boundary between them. By this I mean an alternative to the epistemic division of labour addressed to above, based on the achievement of real transparency over scientific knowledge-production. By real transparency I mean not just making available the scientific or academic papers to the wider public, but by making a huge effort in science communication, in order for lay-people to engage with scientific experts on an equal footing with minimal preparatory effort, should they choose to. It is not sufficient for lay-people to feel scientific knowledge as social or as in any way co-owned by them, by virtue of the scientific papers being available (often at a price) to them. Scientific knowledge for policy-making purposes may be only truly social when lay-people can engage directly with it rather than their interaction being through mediators, both political and scientific.

In order for this to be achieved a sea change is needed in science communication, a move away from the "deficit model" (which I will discuss in later sections) into the perception of co-production and democratisation of scientific knowledge production. The reasons why such a sea-change away from a division of labour is required in the case of science have to do exactly with the importance of science as highlighted in Chapter One. Thus my argument is that because science and scientific knowledge play such an important role in modern societies and scientific knowledge possession is viewed as an extra-celebratory social status, then it is of paramount importance that it should be regarded as a public good and an exemplar of participatory democracy at play.

In the following sections I will scrutinise three series of empirical studies on perceptions of science and scientists in order to find evidence either of the "deficit model" of science communication built into the questionnaires themselves or of the co-production of scientific knowledge. In addressing more specifically the topic of transparency, I will look for indications that lay-people who are interested in science also have sufficient knowledge about the processes of scientific knowledge production, as well as into how they learn about scientific advances both in terms of knowledge production and of new technologies.

2.1.1 Plan

In the first section (Section 2.2) I will briefly engage with three sources of data on lay-people's knowledge of and attitudes towards the institution of science and scientists, by engaging with a series of studies on the topic from the EU, the UK and the US. After presenting data along with a first criticism specific to each one, I will look for evidence of the "deficit model" of science communication at work behind the methodology, and briefly address the question of whether such surveys as construed are genuinely probing on the question of the relationship between scientific expert subgroups and lay-people or whether they make some assumptions that indicate elitist tendencies. Finally I will look for evidence on whether such surveys can after all be informative of the relationship between attitudes towards science and knowledge of scientific processes. I will then turn back to the deficit model, briefly reporting on work by Bauer et al. (2007) that describes and criticises the science communication projects of the last thirty years in terms of three different "deficits", before demonstrating the knowledge deficit model at work through excerpts derived mainly from popular science books. The "deficit model" in science communication will also serve as a demonstration of the lack of transparency over the production of scientific knowledge, as discussed above. I will end the chapter by noting three "techniques" that are used in popular science books and have the effect of transmitting an inaccurate image of the workings of science. Hence, the overall conclusion of this chapter will be that the trust shown by lay-people towards scientists and science as an institution is unwarranted, and that furthermore the lack of transparency (and democratic legitimacy as a result) is there and is cultivated through texts that portray science as an elitist activity.

2.2 Surveys and public perceptions of science

In this section I will review lay perceptions of science in relation to knowledge of science through three series of studies taking place in the EU, UK and US in the last twelve years. I will briefly present the relevant sections of the respective surveys, before seeking to uncover certain methodological assumptions that reinforce the suspicion that such surveys pay lip service to some sort of elitism in how the lay public views scientific subgroups.

The methodology of the studies used varies, with both quantitative and qualitative studies used, as well as a study using mixed methodology (the IPSOS/MORI study regarding the UK)[1].

2.2.1 The EU – Eurobarometer studies

Eurobarometers are the main tool used by the European Commission for gauging public opinion within the area of the European Union and of candidate countries. Their main function is to ensure that the administrative authorities of the European Union really are in touch and consult the citizens of the Union. I will derive data from four Eurobarometer studies having taken place from 2001 until 2010, in which the topic of science was central. These are the following: The Standard Eurobarometer entitled "Europeans, Science and Technology" (Standard Eurobarometer 2001); the Special Eurobarometer "Europeans, Science and Technology" (Special Eurobarometer 2005); the Qualitative Study "The image of science and the research policy of the European Union" (Qualitative Study 2008); and finally the Special Eurobarometer "Science and Technology" (Special Eurobarometer 2010).

2.2.1.1 Summary of results

I will now provide a brief summary of the results of each study, based on the study's own result summary, which comes at the beginning of each

---

[1] There are fierce debates in the social sciences as to the appropriateness and validity of each method, (e.g. Silverman, [2006, ch. 2 and 8]) however in this study I will not dwell any further on such methodological issues.

report.

The avowed aim of the first Eurobarometer study (Standard Eurobarometer 2001, 6) was to gauge European experiences and perceptions of science and technology. The conclusions relevant to my study are the following:

- Europeans regarded themselves as interested in scientific and technological issues, however they thought they were poorly informed on such issues. The main areas of concern were the environment and medicine.
- According to the survey, scientific knowledge has evolved little since the last survey, with the exception of the topic regarding the action of antibiotics on viruses.
- Europeans seem to value and trust scientific professions and research both for the sake of technological benefits as well as simply "advancing scientific knowledge".
- Finally, Europeans believe that scientific and technological practices improve daily life without being a panacea, however they are in favour of social control of scientists by authorities.

The 2005 Special Eurobarometer survey largely confirms the earlier findings regarding high interest in scientific issues accompanied with a significantly lower level of information and active involvement, however the results and analysis are more thorough than previously.

As regards knowledge of science, medicine is regarded as the most "scientific[2]" discipline, with astrology being regarded as more scientific than history and economics, however, when replaced with "horoscopes", the latter comes bottom of the list, suggesting perhaps a conflation of astronomy with astrology. Europeans appear generally quite knowledgeable of scientific facts, scoring highly in quiz questions. Significant socio-cultural and socio-demographic differences arise, however what is striking in this survey is the lack of indicators of knowledge of the processes of science.

Other significant results of this survey are the emphatic agreement to the statement that science and technology make our lives easier, healthier and more comfortable, accompanied however with a more reserved answer (52% agree) to the question whether science's benefits are greater than its harms, and ambivalence towards science as reflected in the 57% agreement to the statement that science and technology are responsible for

---

[2] No distinction is made in the survey between natural sciences and other sciences.

most of the current environmental problems. Finally, more people regard scientific knowledge as important for daily life and a big majority agrees that discoveries in themselves are not good or bad, but it's their use that makes them so.

A related question is posed in another 2005 Special Eurobarometer (2005) concerning decision-making in science, where two in three Europeans agree that decisions about science and technology ought to be taken on the advice of experts concerning risks and benefits rather than the public's analysis. However, no correlation is made between decision-making and public knowledge of the processes of science.

The 2008 Eurobarometer study that I move on to next is a qualitative study, gathering information through free-flowing discussion in focus groups in each EU country. This study reveals a slightly more nuanced perception of what science is, with respondents more often defining science in terms of making progress in knowledge and the explaining of phenomena, as well as in terms of research and the material benefits arising from it. Fewer respondents define science in terms of its methods and rigorous approach. Science is differentiated from technology, which is defined mostly in terms of the products and services that it creates, and from research, which is regarded as a basic component of science. Respondents highly value science and connect it to the idea of progress; however they do express their fears as regards its potential if it remains unchecked.

Finally, a 2010 Special Eurobarometer study confirms the high levels of interest in new scientific findings along with the perceived lack of information on scientific topics observed in earlier surveys, although 91% of the respondents report that they have never attended a public meeting or debate on scientific issues. The findings largely confirm that Europeans tend to think positively of science and technology, however the authors deduce that they do not seem to have a good grasp of the actual daily work done by scientists. This is indirectly deduced from a question on whether the respondents agree that nowadays problems are so complex that they escape the understanding of science and technology specialists; opinions on this question are split, with 37% agreeing, 34% disagreeing and 22% nor agreeing nor disagreeing. As regards decision-making, whereas Europeans feel that scientists should be the ones taking decisions in science, they state that a process of consultation with the broader public ought to take place as well. A final conclusion drawn from this study is the positive correlation between the levels of information about scientific discoveries and positive attitudes towards science and technology. It is notable that whereas the study tests for attitudes towards science and technology, there are no questions directly testing scientific literacy or knowledge.

2.2.1.2 A first criticism

Before proceeding into a first criticism of the Eurobarometer studies, it is worth stating a statistic from them that seems to confirm the suspicion that there is indeed a gulf between the scientific community and lay-people. This is the statistic present in the three quantitative studies that an overwhelming majority of the people probed are interested in science, however many of those respondents also feel that they are poorly informed on scientific topics ([Standard Eurobarometer 2001, 6], [Special Eurobarometer 2005, ch.1], [Special Eurobarometer 2010, 7]). This I take to be revealing of the claim I made earlier, that it may be the case that trust and valuing of science as an institution is not a product of sound knowledge of the processes of scientific knowledge production nor even of the content of the knowledge produced, since Europeans feel poorly informed about science in general. I will now move onto a criticism of the Eurobarometers on the specific topic of scientific knowledge.

Such a criticism of the Eurobarometer studies is the omission of questions on knowledge of the processes surrounding the production of scientific knowledge claims in the latter two Eurobarometers, the 2008 qualitative study and the 2010 Special Eurobarometer, and the shoddy treatment that questions of knowledge receive in the two Eurobarometers (2001 and 2005) where such questions do feature. The omission looks more glaring given that questions of attitudes towards science and scientists are given a prominent role within all four surveys, as well as the admission, at least in the 2005 Eurobarometer survey, that transparency regarding scientific knowledge and technological production is indeed a problem. Hence the 2005 Special Eurobarometer, in its presentation states that:

> It has been noted in the past years that there is a consistent problem of transparency between scientific and technological issues and the information and perception European citizens have on these. It seems that there is an existing gap between science and society: Europeans feel badly informed and little involved in science and technology, which has led to concerns and scepticism on specific issues. (Special Eurobarometer 2005, 3)

Despite flagging up the issue of transparency as a topic deserving careful analysis by means of gauging Europeans' perceptions on it, my claim is that the questions posed only serve to reinforce elitism as regards even the basics of scientific knowledge production. Hence my claim is that as they stand, the Eurobarometer surveys do not genuinely probe into the topic of scientific knowledge as the product of a democratic social activity, nor do they genuinely probe into its baseline characteristics. To reiterate my claim, I consider knowledge of the fundamental processes of science as necessary for transparency as regards scientific knowledge production. I will now turn towards specific criticisms of the Eurobarometers on this topic.

A first criticism is that as regards knowledge of science, even when questions are present in the surveys, they largely confine themselves to quiz questions about purported scientific facts. Even this treatment, however is not well designed: the authors seem happy to treat, in their statistical procedure, questions of long-settled science on the same footing as science-in-the-making questions, asking simply for a true/false answer (see for example [Standard Eurobarometer 2001, 26]). Furthermore, difficult questions relating to the demarcation of science from non-science, value questions regarding the economics of science and its relation to job creation, as well as optimism about curing diseases are being treated in a brief and summary fashion, with the respondents being asked for a simple "inclined to agree/inclined to disagree" answer.

To be fair, the 2001 Eurobarometer does include two questions with the aim of revealing participants' knowledge of what the authors term as "scientific methods", with the topics of the questions being controlled trials on the one hand, and probability as regards genetically inherited disease on the other. I leave aside the issue that the latter question does not seem to correspond to any "scientific method" in order to note that both examples come from the field of medicine, a field which is considered as the key paradigm of a science by most respondents, but however is not accepted as a paradigmatic natural science by most philosophers of science, with physics usually being considered as the main example of a mature natural science.

To return to the topic of indicators of scientific knowledge, the latter Eurobarometer dealing with knowledge of science – the 2005 Special Eurobarometer (Eurobarometer 224) – totally ditches questions of scientific methods, instead opting for a direct question on which disciplines Europeans consider as scientific (with medicine topping the list and the authors trying to explain away why a significant proportion of Europeans regards astrology as scientific) and quiz-type questions in which Europeans are asked for the truth/falsehood of certain statements.

I hence conclude that the Eurobarometers fail to contribute to the solution of the problem of transparency as regards scientific knowledge production, as they do not properly test for knowledge regarding scientific knowledge production.

In order to reach a more considered critique of the lack of study of the relationship between knowledge of scientific knowledge production and perceptions of science, I will now move on to a different survey, scanning perceptions and knowledge of science in the US.

2.2.2 The US – Pew Research survey and Science Indicators

2.2.2.1 Results

I will briefly report on the results of two surveys purporting to report knowledge of science and perceptions of scientists in the US.

I will begin with a July 2009 survey conducted by the Pew Research Center in collaboration with the American Association for the Advancement of Science (AAAS). The survey is interesting in that it scans both the public's attitudes as regards science and scientists as well as perceptions of scientists on their topic and on the scientists/lay-people interface and the various communication problems that arise between the two groups. According to the survey summary, whilst the public rates and respects scientists quite a lot, scientists are critical of the lay public's unrealistic expectations about the speed of scientific achievements and more especially their lack of scientific knowledge, regarding it as a major problem for science ("Public Praises Science; Scientists Fault Public, Media" 2009, 2). Scientists are also highly critical of television and newspaper coverage of science, with 83% and 63% respectively rating TV and newspaper coverage of science as poor or fair (2009, 18). Further regarding the coming together of scientists and lay-people, the authors quizzed scientists on whether they were aware of Science "Town Halls" where scientists and lay-people gather to discuss controversial science questions, with only a quarter of the respondents answering that they had heard "a lot" or "some" about them, even though overwhelming majorities regarded them as at least fairly useful for the public, the news media, policy-makers, and scientists themselves.

As regards knowledge of science, the test is once again one of quiz questions in a multiple-choice answer format, with a 12-question questionnaire being separated into seven "contemporary" questions and five "textbook" ones. Respondents answered around 65% of the questions correctly, being characterised by authors as "knowledgeable" about scientific facts that affect their daily life but less knowledgeable on questions termed as about more complex science topics (2009, 8). On a positive note, however, the phrasing of some of the questions as well as the topics selected should be noted. Questions such as "Which over-the-counter drug do doctors recommend that people take to help prevent heart-attacks?", "According to most astronomers, which of the following is no longer considered a planet?" and "What gas do most scientists believe causes temperatures in the atmosphere to rise?" take seriously into account the role of consensus among the relevant scientific community as regards scientific "facts".

It is worth remarking that the Pew survey devotes a chapter on contentious scientific questions having wider practical and philosophical ramifications, such as the questions of the truth of the theory of evolution and the contribution of humans on global warming (2009, 37-43). A significant question asked along with these questions is the question whether members of the public think that there is general agreement among scientists on the given questions. The authors find marked differences between the percentage of members of the public and scientists who agree that humans have evolved over time, as well as those who agree that the Earth is getting warmer as a result of human activity. Unsurprisingly, the authors find that the views of the public on consensus among scientists are influenced by their own opinion on the topic. However, what is significant for my study is that the question of perceived consensus is being asked at all, something which suggests that the authors want to convey a sophisticated notion of the scientific community among the public.

The second source of US data that I will briefly report on is the 2012 National Science Board Science and Engineering Indicators, an extensive biennial report featuring data from many different sources on a broad range of topics. I will focus on some remarks in the chapter scanning public understanding of and attitudes towards science and technology (National Science Board 2012, ch. 7).

The report includes data from several surveys on factual knowledge of science, concluding that the level of the public's knowledge has not changed significantly over the last two decades (2012, 7-19), and unsurprisingly finding a positive correlation between the level of factual knowledge and level of formal schooling as well as science and mathematics courses completed. The authors recognise the difficulties of measuring factual scientific knowledge over time, given that principles and paradigmatic scientific "facts" are themselves in flux over time (2012, 7-20). The indicators also include a section on reasoning and understanding of scientific processes, with National Science Foundation

(NSF) surveys focussing on probability, experimental design and the scientific method, with the last consisting of an open-ended question on what it means to "study something scientifically" (7-23). It is found that 42% of Americans have a good understanding of scientific inquiry in 2010. Furthermore, respondents aged 65 and over perform significantly lower than younger respondents, and those who have had more formal schooling perform better than those who have not. The study includes as indicators factual knowledge expected from students about science as well as the ability to read charts and statistics, and finally the ability to distinguish science from pseudoscience (with astrology used as a prime example of a pseudoscience).

As regards attitudes towards scientists and science, the indicators show that Americans most trust the relevant expert community to support what is best for the country rather than their professional interests, as opposed to business leaders or elected officials, for a range of topics such as nuclear power and climate change (7-35). Also, deriving data from the University of Chicago Global Social Survey (GSS), the authors find that Americans are ready to grant relevant scientific communities significantly more influence on policy topics such as what to do about global warming and genetically modified foods rather than elected officials or business or religious leaders (7-35).

A final indicator from this survey worth reporting is the degree to which the public believes that there is consensus on a given topical issue among the relevant scientific community. The authors cite a paper by Krosnick et al. (2006) who find, among other things, that "people are more likely to use attitudes and beliefs of which they are certain in forming judgements of national seriousness and policy preferences"(2006, 34) to support the claim that perceived disagreement among scientific experts may be a factor that limits the influence of scientific knowledge and the scientific community on public issues and problems (National Science Board 2012, 32). The relevant indicators show that perceived consensus was highest regarding medical researchers and stem-cell research, followed by nuclear engineers on the risks and benefits of nuclear power and environmental scientists on global warming.

2.2.2.2 A first criticism

The two US studies whose results are reported above seem to be more careful in the wording and the topics chosen on which they pose questions to the public. There are encouraging glimpses of taking into account social and institutional elements of science, even though the NSB seems to ignore earlier calls from within to develop a conceptual model that fully incorporates such elements and considers public knowledge of science as equally consisting of factual scientific knowledge, knowledge of scientific processes and standards, and knowledge of how scientific institutions operate[3].

However, there are some points where a first criticism can be posed. First of all, the data from the Pew Center survey seem to confirm the asymmetry between lay views of scientists and scientists' views of lay-people as regards trust based on mutual appreciation; with elitist assumptions behind such effects as overwhelming majorities of scientists blaming solely the lack of scientific knowledge on the part of the lay-people as the main reason for the bad communication existing between the two communities. Furthermore it seems from the same results that scientists are not active enough in their efforts to engage with and be questioned by the public, even though many admit discussing new research findings with non-scientists ("Public Praises Science; Scientists Fault Public, Media" 2009, 23). A further criticism is that whilst both sources enquire into disagreements among the scientific community on specified topics, no follow-up questions are being asked. For example respondents are not probed for whether they think this is typical of the production of knowledge in scientific settings.

I will now move onto a final source of data on knowledge of and attitudes towards science and scientists, that of a series of IPSOS/MORI polls on public attitudes to science in the UK. As with the other data presented here, I will present data that are significant for my study and that was not present in the previous surveys.

2.2.3 The UK – Public Attitudes to Science series

The "Public Attitudes to Science" surveys consist of a series of four surveys (2000, 2005, 2008 and 2011) commissioned by various departments of the UK government and conducted by different research centres. I will focus on the two later surveys, prepared for the Department for Innovation, Universities and Skills (DIUS) and the Department for Business, Innovation and Skills respectively (BIS), conducted by People Science & Policy/TNS and Ipsos/MORI Social Research Institute.

Methodologically, the UK surveys are the most complex of the ones

[3] The above-mentioned model is a pragmatist model which also considers three main purposes of public knowledge of science, namely civic engagement with science, practical/individual decision-making and cultural curiosity about the scientific world-view (Toumey et al. 2010).

presented so far, as they comprise both quantitative survey methods and qualitative workshops, as well as using a cluster analysis to identify different attitudinal groups in the UK population (People Science and Policy Ltd/TNS 2008, 4).

I will now present briefly some of the novel results of the two surveys before embarking on a first criticism, along the lines of the previous survey data presented.

2.2.3.1 Results

A first key finding is the hierarchy of interest in science, which teases apart aspects that may have been lumped together under "interest in science" in other surveys. Hence, interest for health issues is highest, followed by crime/anti-social behaviour, environmental issues, medical discoveries, music, new inventions and technologies, housing and only then new scientific discoveries as such (2008, 10).

Another interesting source of data is the topic of trust in scientists. The survey data reveal that older respondents, those of lower social grades, readers of tabloids and those with fewer years in formal education seem to be less trusting of scientists (2008, 11). Furthermore, trust in scientists depends on the institutions for which they work, with worries as regards private business funding of science. Finally, for some of the respondents, disagreement in science and the resultant ripe confusion because of it were linked with trust, showing perhaps that for the public it is difficult to distinguish between genuine epistemic disagreement and the defence of interests other than public ones (2008, 11).

The research findings confirmed findings mentioned earlier by US scientists, with respondents identifying lack of scientific knowledge on the part of the public as the main barrier to greater public involvement in science, with lack of interest coming second (2008, 25). This is so, even though the British public feels that, on the one hand they are competent to understand science whilst on the other hand regarding science being "too specialised for most people to understand it" (56% agree vs 24% disagree) (2008, 18). The authors of the 2011 Ipsos/MORI survey note a contrast between similar findings and other data indicating a high level of interest in science (Ipsos MORI 2011a, 51).

Furthermore, most respondents would like "more scientists to spend more time than they do discussing the implications of their research with the general public" and 61% of the respondents feeling that "scientists put too little effort into informing the public about their work" (People Science and Policy Ltd/TNS 2008, 12).

The topic of public consultation with respect to science and technology policy was raised, even though the public appeared sceptical, if not cynical towards public consultations in general (2008, 13). The survey report reveals that the public had little grasp of what public consultation actually consists in, with 25% reporting they did not know what it is and a further 13% giving a vacuous answer (2008, 21). Furthermore, significantly more respondents, when probed, answered that they did not know what the benefits of greater public involvement in decision-making about science, something which the authors interpret either "as scepticism towards public consultation or further evidence of limited public knowledge of what public consultation actually is." (2008, 22)

Even so, the public is quite adamant that the government should act on public concerns about science and technology, disagreeing with the statement that the public is sufficiently involved in decisions about science and technology (48% vs 21% agreeing), and feeling that is important for them personally to be involved in decisions about science and technology[4]. However, they also believe that the government ought to be advised by experts rather than the public on the implications of scientific developments, thus showing a more measured view on the public's input into science policy-making (People Science and Policy Ltd/TNS 2008b, 13). Even though 45% is shown to agree, in the 2011 survey, with the statement that "politicians should put scientific evidence above public opinion when making decisions", (Ipsos MORI 2011a, 54), younger respondents are significantly less likely to favour scientific evidence above public opinion, with those over 75 being significantly more inclined towards scientific evidence. There is also a significant social grade divide, with the more affluent and those with a higher education favouring scientific evidence versus public opinion (2011a, 55).

The public is also probed on the topic of the regulation of science, even though the authors note that from their results it seems that public knowledge of regulation is quite limited (2011a, 32). As regards which bodies should be involved in regulation, on balance respondents thought the government should be a bit less involved and scientists themselves, professional bodies but also the public should be a bit more involved (2011a, 34).

Two final significant results remain to round up the survey data. The first is the probing of knowledge of scientific processes present in the 2011 Ipsos/MORI survey, the second is about the segmentation of the public

[4] However, in the Ipsos/MORI survey, 50% of the respondents agree with the statement "I would like to know that the public are consulted on science issues, but I don't want to be involved personally" (Ipsos MORI 2011a, 54).

into groups based on profiles that the authors take to emerge out of the research. I will describe both in turn.

It seems that when unprompted, the public presents a prototype view of the sciences, with most citing physics, chemistry and biology as the first thing that comes to mind when mentioning the word "science", with fewer citing the medical sciences (2011a, 15). Further prompting revealed that respondents were flexible on what to include as a science, with some insisting on a direct link with the "core sciences" of physics, chemistry and biology, and others mentioning experimentation and the use of mathematics (2011a, 17).

Respondents were further prompted on what the authors termed "Awareness of the Knowledge Production Process", with questions focussing on funding, peer review and the outcomes of scientific research. The choice of these three elements seems to tell us something about the authors' perceptions of the scientific knowledge production process. As regards funding, most respondents thought that most funding comes from the government rather than from industry, (despite the fact that in reality it is the other way round), and also felt that they had very little information on how the topics of research are determined (2011a, 22).

On the topic of peer review, the literature review and the actual survey supply us with conflicting information. The actual survey reports that nearly two-thirds have "a basic understanding of the peer review process"[5] (2011a, 22). A significant minority however does not seem to trust the objectivity of the process, agreeing with the statement that "scientists adjust their findings to get the answers they want" (2011a, 22). On the other hand, the literature review cites evidence from a 2003 Ipsos/MORI survey according to which 75% of the British public did not know what "peer review" meant or could not define it correctly, and from a 2002 Ipsos/MORI survey showing that 71% of the public expected scientists to give an "agreed view" on scientific issues (Ipsos MORI 2011b, 4). The notion of agreement by scientists is also linked to trust in scientific information, with the authors devoting a whole section (Section 4.1) on the topic. A suggestion drawn from this part of the survey by the authors is that "there is potential to raise trust in science by raising awareness of peer review" (Ipsos MORI 2011a, 38).

A final new contribution that the PAS surveys make towards the understanding of the public's perceptions of and attitudes towards science is the segmentation of the public into distinct profiles, based on a cluster

[5] Based on agreement with the statement that "before scientific findings are announced, other scientists have checked them"

analysis of the final data from the quantitative survey. The authors devote a whole chapter (2011a, Ch.8) on a detailed analysis of the six profile clusters[6], studying their defining characteristics, demographic make-up and other features, as well as the implications of the analysis as regards suggestions for engagement on science and science issues. However, they do state the proviso that the profiles are based on similarity of attitudes across a wide area of the questionnaires, rather than identity of answers and hence the people falling under a given profile are likely to hold a certain view, rather than actually possess that view.

2.2.3.2 A first criticism

The PAS studies seem to be quite a step up from the previous studies in terms of thoroughness and diversity of methods used. It should be noted that the correlation between education levels and trust in scientists is explicitly probed for, however this refers to general level of formal education, rather than education specifically in the sciences. There is further evidence to support a negative answer to my guiding question of trust as correlated with knowledge of the processes of scientific knowledge production, with a significant amount of respondents not having sufficient knowledge to understand what "peer review" is, along with a significant portion of the population that does not trust that scientific processes are totally objective.

As regards transparency, a clear answer is given by the respondents, most of which regard lack of scientific knowledge as the main barrier for more public involvement in science. This stance confirms the findings of the other two studies. The survey also yields significant amount of data on the desirability of public involvement in science, with younger respondents favouring public opinion over expert scientific opinions as regards policy-making. A further statistic which is revealing of the complexities of opinion mining, and which has been ignored by the Eurobarometer and the Pew studies is the social stratification of the respondents, with more affluent respondents favouring expert opinion as opposed to less affluent ones who favour public opinion as an input to political decision-making.

A few minor criticisms of the study are also in order. Difficulties in the interpretation of the questions posed surface in the report, with the authors

---

[6] The six profile cluster labels are the following: the Concerned (around 23% of respondents), the Indifferent (19%), the Late Adopters (18%), the Confident Engagers (14%), the Distrustful Engagers (13%) and finally the Disengaged Sceptics (13%).

worrying that when they asked respondents what their worries were about research on a specific issue such as climate change, the question was often interpreted as how worried the respondents were about the issue itself. The authors recognise this, for this specific topic (People Science and Policy Ltd/TNS 2008a, 28), however this phenomenon of an interpretation of questions different than the one the authors of the study intend, especially in quantitative research, may be more endemic than noted.

2.2.4 A more general criticism of the surveys and their methodologies

In this section I will review some more general criticisms of the surveys' methodologies, in order to make the reader aware of their limitations as regards their validity.

An example of equivocation that may be less innocuous than it first appears can be drawn from an indicative question found in the 2010 Eurobarometer. The section focusses on science communication, with respondents asked which category of people is better qualified to explain the impact of scientific and technological developments on society (Special Eurobarometer 2010, 90). The latter question seems to contain an element of cunningness, by introducing the element of qualifications, which may explain the high preference for scientists, given that scientists are probably the ones with the most formal qualifications, assuming that no formal qualifications are given for moral or social responsibility.

Another source of criticism concerning the methodology of qualitative and quantitative research and the validity of the results obtained is in order. I will briefly mention two worries, one for each methodology. The first source of worries is about qualitative studies, with two main issues arising; first, regarding the external validity of the results obtained, that is the question whether the opinions voiced during the interview are representative of opinions held by the wider population; the second is whether the respondents' stance reflects their own stance or has come about as a result of the general opinions voiced during the discussions. That is, whether the respondents would give the same answers if interviewed separately, or whether their answers were more a reflection of the group stance rather than their own.

A second general worry concerns the validity of questionnaire results and is akin to the placebo effect, in the sense that the respondents may answer questions in a way that they believe would please the researchers or that is socially acceptable. The authors of the 2010 Eurobarometer study recognise this pitfall in a footnote (Special Eurobarometer 2010, 9) and hence are quite measured in drawing conclusions about Europeans' interest in scientific and technological issues. The above-mentioned possible source of bias is well-known in the literature, where it is labelled as socially desirable responding. According to Paulhus, socially desirable responding is the tendency to "give answers that make the respondent look good" (Paulhus 1991, 17). Phillips (1973, ch. 3) distinguishes between and tests two measures of socially desirable responding, what he terms as "trait desirability" and "need for approval". The first, trait desirability, refers to the hypothesis that "the more desirable people's assessment of the traits, the greater the extent to which they will report" them, whilst the second refers to the notion that "the greater people's need for social approval, the greater the extent to which they will report" them (1973, 43). Both varieties of social desirability bias may be at work here, especially the need for social approval, as regards especially questions about the potential powers of science and whether they make life easier, questions in which the socially desirable answer is quite obvious. However, the problem is that even using methods of acknowledging the effects of socially desirable responding, the separation of the real element (the response to the content of the question) and the biasing element is often a very difficult matter.

A further source of bias is what is labelled as acquiescence bias (1973, ch.2), described variously by its two varieties, "yeaysaying" and "naysaying" (Couch and Keniston 1960). "Yeaysaying" refers to agreement with the item in question irrespective of its content whilst "naysaying" refers to disagreement with the item irrespective of its content. According to Peabody ([Peabody 1966] mentioned in [Paulhus 1991, 47]), the tendency of acquiescence is not to be understood as a mechanical response to any question, but rather it is assumed to emerge when the subject is uncertain about the question.

The two response biases studies above pose a threat to the validity of the results, especially given that there exist at least some studies ([Phillips 1973], [Dohrenwend 1966]) that claim that biasing effects may vary non-randomly across different profiles, such as those yielded by socioeconomic data. Hence, the validity of the PAS study classification may be threatened by such biasing effects.

A final element of equivocation is to be found on the section of the 2010 Eurobarometer questionnaire that focusses on scientists and society. The section focusses on science communication, with respondents asked to state their agreement or otherwise on whether "Scientists do not put enough effort into informing the public about new developments in science and technology" (Special Eurobarometer 2010, 89) and being asked which category of people is better qualified to explain the impact of scientific and technological developments on society (2010, 90). The latter question, as mentioned above, introduces the element of formal qualifications, something which is mostly associated with scientists. However, no formal qualifications are awarded for moral or social responsibility.

2.2.5 Data interpretation

A crucial question that needs to be asked at this point is whether, given the criticisms posed for each of the three studies, the studies actually do what they purport to do, that is give an accurate picture of lay perceptions of scientists and the production of knowledge in scientific settings, or whether they are merely reflecting their authors' or funders' biases or some idealised but unrealistic views of science held by government agencies.

A first criticism is to point out what Phillips (1973, ch. 4) calls modelling effects in his exploration of possible sources of bias in sociological surveys. The effect, according to Phillips, occurs when "the investigator consciously or unconsciously projects his own views (attitudes, opinions, or whatever) on those whom he studies". (1973, 60). Phillips tests for this effect in his study, his results lending support to the suggestion[7] that the respondent may attempt to answer questions in a manner that agrees with their perception of the interviewer's opinion in order to maintain or even increase the reward value of the interaction. Phillips hypothesises that in face-to-face communication such effects are increased, even unwittingly, given the fact that there are many extralinguistic cues available, and hence such a bias may also contaminate findings (1973, 67).

A second criticism that may be revealing either of the difficulties of gauging lay knowledge of science or of the authors using simplifying assumptions about it is the way "scientific knowledge" is tested for, namely through quiz questions, usually about "textbook" facts, and more rarely probing into things such as "peer review" (Ipsos MORI 2011a) or using careful wording that would bring forth the social nature of scientific knowledge production. It is illuminating at this point to look into a threefold distinction that John Durant makes on the topic of scientific literacy. Durant distinguishes between "knowing a lot of science", "knowing how science works", and "knowing how science really works" ([Durant 1993] cited in [Gregory and Miller 2000, 89]). By this distinction

[7] Attributed by Phillips to Williams (1969, pg. 26)

Durant refers to knowledge of scientific facts, knowledge of methodological and other formal criteria of science ideally construed, and finally of the perception of science as a social activity performed within communities of practitioners who often have particular professional and personal interests. Durant claims that in terms of scientific literacy, the first definition, that of knowing many scientific facts, is at best necessary but not sufficient as an indicator of scientific literacy (Durant 1993, 134)[8].

To begin moving towards a conclusion, it seems that there are good reasons to support the notion that, given the sources of bias and the shoddy treatment that questions of knowledge of the production of knowledge by scientists, the above-mentioned surveys are not at all revealing of the actual perceptions of science by lay-people. If this is the case, then the survey data may be seen as merely an attempt to pay lip service to government and scientific bodies seeking legitimation through the authority of science. To anticipate more a thread which I will return to later regarding the relationship between science and policy, it may be claimed that the survey data serve the purpose of the mutual legitimation between science and state institutions (Funtowicz and Strand 2007, 8). More specifically, the data would be revealing of the "deficit model" of the public understanding of science, in which the public is "deficient" in some quantity (knowledge, attitudes or trust) which scientists possess and may transfer to them. However, I will elaborate more on the deficit model in the following sections.

This conclusion nonetheless strikes me as a bit too quick. True, biases such as the social desirability bias may be seen to threaten the validity of such data as the profiling performed in the PAS studies, however I feel compelled to argue that through their diversity the surveys do reveal certain general opinions held by lay-people as regards scientists. Hence, it does seem that people value science and scientists but at the same time, this valuing does not seem to rest on a well-founded knowledge of the processes and social dimensions of science. Furthermore, the public seems to discern a relationship between science and policy-making. This is so even if through the questionnaires it is not extensively probed as to where the public's own role may lie in decision-making. The public also seems to demand some degree of accountability and responsibility from scientists, as well as more of a say into what topics get to be researched. Finally, the

[8] The reason Durant cites is that most technical and scientific issues that citizens are confronted with in their daily lives regard cases of science-in-the-making rather than settled (hence not contested) science, and, as Latour poignantly argues, (Latour 1987) newly produced scientific knowledge is more often than not controversial and contested, as opposed to long- established scientific knowledge.

lay public does seem to value the institution of science, and does seem keen to both learn more and appear to be more informed about it. In light of the above, I conclude that the data from the above surveys should be treated with caution, but should not be totally disregarded. As regards answering the two guiding questions that I laid down above, that of transparency and that of unwarranted trust, I conclude that the questionnaires do provide answers to them, even though unwittingly, without explicit probing into them (even though there are questions in all three series of questionnaires that gesture towards an answer, they are not addressed thoroughly and explicitly). At the same time, the correlation of ignorance of the processes and social dimensions of scientific knowledge production ties in with the alleged elitism of the scientific "class". Unsurprisingly, evidence of this elitism is to be found unwittingly in the choice of questions posed in the questionnaires, as well as the methodology used in order to obtain the results. As evidence of that, it can be discerned that the methodology itself seems to not do justice to the opinions of lay-people, in the sense that what may be quite sophisticated responses on the part of lay-people are often blocked by the questionnaire questions which simply ask for degrees of agreement at best and for simple true/false or agree/disagree answers at worst. This may be claimed to be an unfortunate consequence of the need to gauge an indication of the public opinion, however it should be recognised that this is a simplification of the public opinion, which may be infinitely more complex and sophisticated. Not recognising the above may open one into charges of elitism on the basis that they may hold the opinion that the lay public does not possess sophisticated opinions as regards topics on the intersection of science and the public sphere, as opposed to themselves. As Ravetz notes,

> In spite of professions of democratic sentiment, science is part of elite culture. The very language of science, explicit, logical, formalised, and technically esoteric, requires a style of thinking that is almost totally restricted to those with a lengthy (and expensive) education. It contrasts with the informal, partly tacit, situated and anecdotal knowledge used by those less favoured people who actually keep our system running (as in the 'Murphy's Law' literature) (Ravetz 2011, 144)

There are plenty of instances in the surveys of the clash between what Ravetz calls the formalised language of science and the informal knowledge of the lay public, especially in the qualitative studies where opinions of lay-people are transcribed verbatim.

In the following section then, I will elaborate on how the above charge of elitism gets realised through the "deficit model" of the public understanding of science.

2.3 The "deficit model" of the public understanding of science

In this section I will elaborate a bit on what scholars in several related fields such as Science and Technology Studies, the Public Understanding of Science and Science Communication have termed "the deficit model" of the public understanding of science. Actually, based on a review by Bauer et al. (Bauer, Allum, and Miller 2007), I will present three types of assumed "deficit". The three types are given as related to a history of efforts at communicating science to the lay public, and my criticisms will closely reflect those of the authors of the review, as well as build on the criticisms levelled at the survey studies. This will pave the way for a study of the effects of the various deficit models on popular science and their impact on the perceptions of science by lay-people.

2.3.1 General Overview

Deficit model paradigms are built on a separation between two communities, those of the experts and the lay public. The metaphor which most accurately describes the deficit models is that of one group's minds being akin to a bucket devoid or deficient in some sort of substance, and the other group possessing that substance and pouring it into the minds of the deficient group until they are full (see Gregory and Miller [2000, 244–250] for the bucket analogy). What is transferred and which group is the deficient one differs from group to group; in the first paradigm it is knowledge, in the second it is attitudes, in the third it is trust. I will now describe the three main paradigms using deficit models, associated by Bauer et al. (2007) with bridging the gap between the experts and the lay-people on the one hand, and with the history of academic and policy research towards this end on the other.

2.3.2 The Scientific Literacy paradigm

The Scientific Literacy paradigm was the main paradigm driving research into bridging the perceived gap between science and society from the time when this became a problem in the mid-1960s till the mid-1980s[9]

[9] The dates cited indicate when the paradigm was dominant. This does not mean

(Bauer et al., 2007, 80). According to this paradigm, the deficit is one of knowledge. The driving idea was that the lay public simply did not know enough facts about science, with the assumption being that if the public improved its stock of scientific facts stored in their minds, then they would obviously be more positively disposed towards science. Furthermore, this model disqualifies lay-people from technocratic decision-making; since lay-people are ignorant on certain technical matters pertaining to policy-making their opinion carries no weight at all (2007, 80).

A naïve deficit model would revolve around factual knowledge of science or of facts produced in scientific settings, hence subscribing to the first part of the typology laid down by Durant (1993, 131) and described above as "knowing a lot of science". However, more sophisticated notions of scientific literacy take the concept to include the following four elements (cited in Bauer et al. (2007, 80-81)) :

- knowledge of textbook facts of science
- an understanding of scientific methods such as probability reasoning and experimental design
- an appreciation of the positive outcomes of science and technology for science and
- the rejection of superstitious beliefs such as astrology and numerology.

A criticism of this typology is that it presupposes that to be scientifically literate automatically means to be positively disposed towards science. This is coded more on the normative elements described in the last two elements of Bauer et al.'s typology. It seems like a very simple-minded way to see the world, to regard scientific literacy as a virtue, and at the same time demand that any scientifically literate person, in order to qualify as such, would have to be positively disposed towards science. Furthermore, it seems that none of the above criteria is realistic, for the following reasons.

The criterion of knowing a lot of textbook facts about science may break down once one moves beyond the textbooks, in the sense that often textbook facts are viewed as only approximately true by experts, if not as outright false[10]. Hence consultation with experts may actually result in contradictions arising between the expert opinion on a given question and

---

that elements of the paradigm are not in use even today. Indeed, elements of the deficit in knowledge paradigm are to be found in perhaps all three of the Eurobarometer studies.

[10] Especially if no good account of approximate truth can be provided, and one is left with a two-valued truth semantics.

the textbook one. Furthermore, the insistence on textbook facts glosses over the importance of knowledge of science-in-the-making.

The criterion of knowledge of scientific methods is a controversial criterion, especially given the diversity of methods used in what people qualify as science, even when this is restricted to the natural sciences. Furthermore, the relationship between the methods of science and common sense is quite hard to elucidate and to measure, in the sense that a method, if labelled as (say) probabilistic thinking and formalised, may seem to be remote and inaccessible to a lay-person, whereas that lay-person may be an expert in manipulating things close to their lives involving probabilities, such as horse-racing odds. A key disadvantage and beginning of many attacks on deficit models is the lack of contextualisation, something which the example just given brings out.

The appreciation of the positive outcomes of science and technology for science also seems to be a very shaky criterion, especially because it involves an explicit evaluation but also because it is an all-or-nothing notion, in the sense that a different answer regarding the impact of science would be given by a student researching CT-scan imaging algorithms from a person living in the Chernobyl area and suffering from a horrific cancer[11] , whereas perhaps both need to be taken into account.

Finally, the rejection of superstitious beliefs is also a spurious and unrealistic element, as what counts as scientific belief as opposed to superstitious belief is often highly culturally-dependent. For example, one probably assumes that Einstein, when declaring that "God does not play dice with the Universe" was not thinking in a superstitious way[12] but rather in a metaphorical one, however research that follows all canons of scientific method but attempts to vindicate practices like parapsychology is often not accepted or not even allowed to get off the ground (see for example [Collins 1985]).

In all, it seems that the scientific literacy deficit model is misguided at best and pernicious at worst. However it has a strong grip over what it means to be familiar with science, to the point where surveys using questions such as the Eurobarometers and the Pew Center ones – questions on whether respondents agreed or not with textbook examples – seem to be geared towards vindicating this deficit model, and policy is formulated

---

[11] It should be obvious that such example may be produced ad nauseam, especially if one considers issues such as the effects of climate change or of modernisation on indigenous populations.

[12] Nor was Newton in his forays into alchemy or theology, which have to be squared with his purported claim that "hypotheses non fingo" (though see [Cohen 1962] disputing the claim)

taking such evidence as good evidence of an existing problem.

I will now move on to a second deficit-based paradigm, that of the public understanding of science.

2.3.3 The public understanding of science (PUS)

The public understanding of science paradigm had its heyday, according to Bauer et al. (2007, 82) from 1985 until the mid-nineties. The lay public was again the deficient party, however, this time, it was deficient primarily not in knowledge but in positive attitudes towards science. The danger is not that citizens will be ignorant about science, but rather will be neutral or even actively anti-science. This, Bauer et al. note, would of course be bad for scientific institutions and for institutions that fund and depend on science, such as the state.[13]

What is worth noticing is that a dominant assumption of the scientific literacy paradigm carried over to the public understanding of science model: it was assumed that the more people would get to know about science, the more they would be more favourably disposed towards it. However, the issue now was no longer all or nothing: as opposed to being scientifically literate or not, the lay public could be listed on a continuum of being more or less knowledgeable (2007, 83).

Bauer et al. attribute two separate agendas to the PUS paradigm, the normative-rationalist agenda and the realist-empiricist agenda. The normative-rationalist agenda is based on the assumption that negative attitudes and biases are the products of a lack of knowledge and rationality; hence the public ought to be educated into thinking rationally, so that it will produce dispassionate and rational judgements, in the same way that purportedly experts do. On the other hand, the realist-empiricist agenda takes attitudes to be primitive and just out there in the world and attempts to directly swing them in favour of science - in a way scientists should try to win "hearts and minds". In such a case the lay public are treated as consumers of science, and marketing techniques are used to woo them (2007, 83).

Not surprisingly, a main line of criticism of the PUS paradigm is to attack the notion that the more one knows science the more they would be favourably disposed towards it it. Bauer et al. give ample evidence that this need not be the case: they cite survey studies which seem to indicate

---

[13] It may not be a total coincidence that the Superconducting Super Collider (SSC) project was cancelled in 1993. This had a significant effect on at least the community of particle physicists and their lobby, and perhaps, as a result, to the whole domain of public understanding of science.

that for a non-negligible number of people, familiarity with science either breeds utter contempt or simply indifference. As instances of the dubious links between knowledge of science and specifically positive attitudes the authors cite research in attitude theory which reports that "knowledge makes the difference between attitudes and non-attitudes, and not between positive or negative attitudes". ([Converse, 1964] cited in Bauer et al. (2007, 84))

The authors also cite research conducted by Sturgis and Allum (2004) that showed that the level of "general political sophistication" was connected to more positive attitudes towards science.

Bauer et al. propose that such a deficit model becomes engrained in a vicious circle of mistrust between the public and the experts, in the sense that mistrust on the part of the experts towards the public and mistrust on the part of the public towards the experts reinforce each other. In such a case, an analysis in terms of attitudes and an attempt to change these attitudes, if made explicitly, is probably doomed to failure, given that if it is recognised for what it is, it may further stoke mistrust among the public.

As regards the incorporation of the insights of this model in the survey studies, a shift from a heavy focus on knowledge to one more geared towards measuring attitudes is evident in the comparison between (for example) the topics addressed in the Eurobarometer 2001 and 2010 quantitative studies, with the latter containing a large chapter on attitudes towards science. I will now move onto the last of the three deficit models, that of the deficit of trust, associated with the science and society paradigm.

2.3.4 Science and Society

A final deficit-model based paradigm has been dominant, according to Bauer et al., from the mid-nineties until the present. In this model, it is not the public but rather experts and policy-makers that are deficient, and the quality they are deficient in is trust. Hence, in other words, the experts and policy-makers ought to do work to gain or regain the public's trust. The idea is that false conceptions of the public by scientific experts lead to misguided policies regarding communication and these policies serve to further alienate the public (2007, 85).

According to the authors, the shift towards the science and society paradigm has presented a quite significant break from the earlier models. In this paradigm, analysis and intervention on the part of the experts is no longer separated, but rather the aim of academic analysis is to change institutions and policy. Furthermore, openness and consultation with the public even at the early stages of scientific research and policy for science is desired. Hence, the aim is for citizens to be involved at every stage of scientific development and the genesis of scientific knowledge, rather than be presented with ready-made facts. Modes of active participation are sought after, and practices such as consensus conferencing, national debates, citizen juries etc., have been gaining in importance and visibility on many levels in the last ten years.

The authors are critical of the honesty of the promoters of such an agenda of increasing public participation in science,[14] claiming that what they are after is not genuine co-production of knowledge and policy, but rather the aim is to maintain the older deficit models through massive exercises of public persuasion. They however cite the failure of the British government to persuade the people about genetically modified foods during the GM Nation public debate and the ensuing reactions on the government side as support for their suspicions; according to them, the responses were either to attack the process itself for allowing environmental groups too much influence over the wider public, or to propose that further dialogue was needed until the public had the "right" attitude towards genetically modified crops. Hence it seems that what the government was after was not genuine and honest dialogue, but rather a situation where it would be doing most of the talking and probing and the public would only be listening and nodding their head in agreement.

Going back to the material discussed in the previous chapter, the "science and society" agenda seems to be in principle a good step towards the democratisation of knowledge production in scientific settings. In practice, however, and if one accepts the criticism of the authors, the government seems to not be seeking genuine participation but merely legitimation and agreement by an understanding, but uncritical public. Hence, I conclude, it is far from the notions of democratic deliberation and participatory democracy, as described in Chapter One. Furthermore, there is evidence, also to be found in the 2010 Eurobarometer that briefly touches on these topics, that what the EU is interested in is less than the democratisation and co-production of scientific knowledge, despite the increase in spending in Science Communication projects (see for example the CASC project recommendations, [Harper 2011]).

In the following section, I will mainly focus on shining some light on how the elitist idea of a deficit in knowledge is implemented mainly through popular science writings that often shape attitudes towards science.

---

[14] Incidentally, they regard public consultation beyond the parliamentary process as symptomatic of a wider policy on the part of the government, that is not restricted to science policy.

2.4 Deficits and the shaping of elitist perceptions of science

In the sections above I have demonstrated, through the study of the survey studies, that there appears to be a lack of transparency over the production of knowledge in scientific settings, and that furthermore the trust that lay-people show to scientists and to scientific institutions is largely not based upon a sound knowledge of the processes by which scientific knowledge is produced. I have further argued that science communication operates in an elitist fashion, by using deficit models that conceptualise the public as deficient and of scientists as possessors of a special "substance" which they transmit to the lay public. In this section I will give some examples of how in practice the asymmetrical relationship between scientists and lay public is maintained, through the citation of techniques used by science popularisers to propagate an elitism about science.

For the moment I will focus on substantiating my claims through an exposure of enthusiastic and often sensationalist portrayals of scientist-heroes or of science as the pinnacle of human intellectual activity or in any case as a treasure-trove of facts deserving to fill the lay-person's mind, along with some brief comments for each, before moving on to the second section where I will attempt to discern some general characteristics and some ways by which such an unfaithful portrayal may be mitigated.

2.4.1 Waxing Lyrical about science and scientists

The first extracts that I will cite come from a book entitled "*The new optimists*" (Richards 2010) and subtitled "scientists view tomorrow's world & what it means to us". The book would fall in the "popular science book", whilst its publisher, Linus Publishing, is a not-for-profit organisation "dedicated to producing high quality educational materials including textbooks, anthologies and lab manuals" (Linus Publications 2009). Hence, this book is explicitly an effort aimed at educating the lay public about science. I will not list passages from the content of the book itself, rather I will cite passages from the back cover and from the foreword.

The back cover passage finishes with the following paragraph:

> **The New Optimists** provides a spellbinding insight into the minds of some of the UK's leading scientists and shows just how much work is being undertaken on our behalf in some of the country's leading institutions. It will open your mind to the endless possibilities of focused and responsible research – and help us all face the future with confidence. (Richards, 2010)

In this passage we can notice the appeal to the brilliance of mind of leading scientists, as well as the overtly optimistic language of the text. Furthermore, in the foreword we may notice the building up of a mythology affirmative of scientific genius:

> In 1660, the men who gathered at Gresham College in London, determined to explore the universe they inhabited, decided to form a society to enjoy, said their historian Thomas Sprat, "the satisfaction of breathing a freer air, and of conversation in quiet with one with another, without being engaged in the passions, and madness of that dismal age". In Birmingham, that spirit was continued a century later in the men of the Lunar Society, among them James Watt and Matthew Boulton, Erasmus Darwin, Josiah Wedgwood, Joseph Priestley and the chemist James Keir. Their interests ranged from astronomy, optics and electricity to chemistry, engineering and metallurgy, and to medicine and botany, and their long collaboration brought together their different experiences and skills, providing lifelong support. They too looked to the future, setting aside political differences, and concerned above all to make the world a better place. (Jenny Uglow, foreword in Richards [2010, v])

Here we notice, in the quotation of Thomas Sprat, the notion of the scientists battling, in his world, "the madness of that dismal age"[15], as well as the notion that one of the main motivations of the scientists was to work for a brighter future for humanity.

I will now move on to some excerpts of high-profile books popularising science, written by top scientists. The first excerpt that I will look into concerns a prominent scientist and populariser of two arcane branches of physics, cosmology and string theory. Brian Greene hence recounts how as a teenager he became infatuated with the wonders of the cosmos, whilst at the same time acknowledging an author and philosopher, Albert Camus, as one of his main influences:

> I was struck by Camus' ability to discern hope where most others would see only despair. But as a teenager, and only more so in the decades since, I found that I couldn't embrace Camus' assertion that a deeper understanding of the universe would fail to make life more rich or worthwhile. Whereas Sisyphus was Camus' hero, the greatest of scientists – Newton, Einstein, Niels Bohr, and Richard Feynman – became mine. And when I read Feynman's description of a rose – in which he explained how he could experience the fragrance and beauty of the flower as fully as anyone, but how his knowledge of physics enriched the experience enormously

[15] Sprat is referring to the Civil War that was gripping England at the time of writing.

> because he could also take in the wonder and magnificence of the underlying molecular, atomic, and subatomic processes – I was hooked for good. I wanted what Feynman described: to assess life and to experience the universe on all possible levels, not just those that happened to be accessible to our frail human senses. The search for the deepest understanding of the cosmos became my lifeblood. (Greene 2004, 21)

In this excerpt we may see the reference to heroic scientist-figures of the past, as well as the description of almost mystical experiences and a quest for ultimate reality, couched in almost religious terms.

The next excerpt comes from a self-proclaimed defender of science on a mission to debunk muddled thinking and superstition. Alan Sokal became an eminent figure in the so-called "Science Wars" of the mid-nineties, with his article "Transgressing the Boundaries: Towards a Transformative Hermeneutics of Quantum Gravity" and its subsequent revelation as a hoax playing an important role ridiculing vast swathes of academia[16]. In his Preface, Sokal explains why he wrote the book thus:

> At a superficial level the topic is the relation between science and society; but the deeper theme is the importance, not so much of *science*[17] but of the *scientific worldview* – a concept that I shall define more precisely in successive chapters, and which is in no way limited to the natural sciences – in humanity's collective decision-making. Whether my targets are the postmodernists of the left, the fundamentalists of the right, or the muddle-headed of all political and apolitical stripes, my refrain is the same: clear thinking, combined with a respect for *evidence* – especially inconvenient and unwanted evidence, evidence that challenges our preconceptions – are of the utmost importance to the survival of the human race in the twenty-first century.(Sokal 2008, xi)

Sokal continues:

> This book belongs to a fairly rare genre: that of a natural scientist writing for the general educated public on cultural issues that are only indirectly related to his field of research and teaching. (2008, xi)

Two main lines of brief criticism are in order: first, the notion that clear thinking is only the province of the "scientific worldview" which Sokal goes on to define. It often seems that Sokal wants to claim both that science just is good common sense, in which case the purported

[16] Unfortunately, subsequent reverse hoaxes published in scientific journals (such as the Bogdanov affair - (Wikipedia 2012c) became much less notorious.

[17] The italics are to be found in the original.

worldview need not be labelled scientific, or that there is indeed something special which is the sole property of the thought processes of scientists from the Scientific Revolution onwards; and that this provides us with a privileged access to reality. Clearly, in regarding postmodernist and religious thinking as muddled and as prominent so as to constitute a danger to the survival of humankind, Sokal seems to favour the latter disjunct, in which case he has to importantly substantiate his definition of the scientific worldview and differentiate it from good common sense.[18]

The second criticism concerns the self-proclaimed mission of the book. Sokal takes himself to be representative of a community under siege, that of scientists deliberating about public issues that are not within their expertise. However, this does not seem to be the case, especially given that scientists are often asked to do exactly that, namely to suggest policy, sometimes on issues that are more cultural than technical, or in any case outside of their field of expertise (such as natural scientists grappling with economic issues). Furthermore, when it is pointed out to scientists that they are deliberating on issues that are not within their reach, such as in recent cases in the UK over scientists deliberating on a cultural practice such as homeopathy or drug taking, then there is sometimes huge uproar about "freedom of speech".[19]

A final acclamation about a particular scientific theory comes from Richard Dawkins' Introduction to the 1996 edition of his popular science book "*The Blind Watchmaker*" (1996, x)

> Darwinism encompasses all of life – human, animal, plant, bacterial, and, if I am right in the last chapter of this book, extraterrestrial. It provides the only satisfying explanation for why we all exist, why we are the way that we are. It is the bedrock on which rest all the disciplines known as the humanities. I do not mean that history, literary criticism, and the law should be recast in a specifically Darwinian mould. Far from it, very far. But all human works are the products of brains, brains are evolved data processing devices, and we shall misunderstand their works if we forget this fundamental fact. If more doctors understood Darwinism, humanity would not now be facing a crisis of antibiotic resistance. Darwinian evolution, as one reviewer has observed, "is the most portentous natural truth that science has yet discovered". I'd add, "or is likely to discover".

---

[18] I assume here that humans have always, in their own ways, paid attention to evidence and that most, if not all, humans are endowed at least with the capacity for clear thinking.

[19] Such as the David Nutt affair for example, see his letter to the Times dated November 2nd, 2009 (Nutt 2009).

In this excerpt, one may notice the exaggerated language in describing the scientist's favourite theory, and the perception of its significance couched in quasi-religious terms[20]. Furthermore, there is a recommendation that areas other than the scientist's expertise follow its lead and recognise that they are ultimately based on it. The prescription seems to be especially audacious as it covers both the cultural sciences, a favourite target of enthusiastic scientists, but also medicine, a step that few enthusiasts seem ready to make.

More enthusiasm is shown in the two opening paragraphs of the preface:

> This book is written in the conviction that our own existence once presented the greatest of all mysteries, but that it is a mystery no longer because it is solved. Darwin and Wallace solved it, though we shall continue to add footnotes to their solution for a while yet. I wrote the book because I was surprised that so many people seemed not only unaware of the elegant and beautiful solution to this deepest of problems but, incredibly, in many cases actually unaware that there was a problem in the first place! The problem is that of complex design. (...) The complexity of living organisms is matched by the elegant efficiency of their apparent design. If anyone doesn't agree that this amount of complex design cries out for an explanation, I give up. No, on second thoughts I don't give up, because one of my aims in the book is to convey something of the sheer wonder of biological complexity to those whose eyes have not been opened to it. But having built up the mystery, my other main aim is to remove it again by explaining the solution. (Dawkins 1996, xv)

Here the author attempts to overwhelm the reader with quick-fire scientific facts, whereas they also seem attempt to portray those who do not see the objects of their enquiry as complex as ignoramuses in dire need of education. Hence, the deficit is very real and the author is attempting to fill it in with their knowledge.

A final excerpt in a more timid tone, also reveals a perceived deficit of knowledge, to be addressed by a bombardment of quick-fire scientific facts. The excerpt is from Steve Jones' book "*Y: The descent of men*", who uses a very entertaining style in stark contrast with Dawkins, but nonetheless uses the same device of a bombardment of facts that may sometimes mask the subtlety of the argument presented:

> Sperm evolved to carry genes but is now used to move foreign DNA into

[20] It is perhaps an irony that, at least Orthodox Christian theology, is couched in terms of portentous truths.

> eggs (the main success so far has been to make mice that glow in the dark). Human proteins are made in animal semen, which turns the litre of ejaculate produced by a boar into a useful commodity. Man himself may in the end become redundant, for his sperm can be grown in animal testes, and in mice at least an egg can be fertilised with a body cell from another female, which cuts out the secondary sex altogether. (Jones 2003, 7)

Here again, the author presents a multitude of scientific facts in a few phrases, without providing the reader with any possibility of cross-examination or criticism. Hence the idea of the communication of science as filling a deficit of scientific facts within the lay-person's mind is again seen at work.

2.5 Conclusion: Some general remarks on deficits and public perceptions of science

I started the current chapter by posing two questions, that of transparency of scientific knowledge production and that of trust in science in correlation with a sound knowledge of scientific knowledge production. I proceeded in looking for an answer through surveys in the EU, US and UK. I criticised the survey methodology as revealing of elitist biases, but concluded that it can indeed be informative of lay understanding and attitudes towards science. Having identified some elitist assumptions at work, I then presented work by Bauer et al. (2007), who analyse three separate research programmes in the communication of science in terms of three different sorts of deficits.

Finally, I gave some concrete examples from popular science books, written by prominent scientists, which, according to my argument, serve to propagate elitist conceptions of science. In the final section of the chapter I will elucidate a bit more what I regard as the mechanisms by which science popularisers propagate, perhaps unwittingly through their writings, an elitist conception of science.

2.5.1 Settled science versus science-in-the-making

The first topic that I will explore is the topic of settled science versus science-in-the-making and of popular scientific domains. The examples of books given above, or more generally the popular science books that usually attract a big number of readers, are about arcane areas of science that are surrounded by an air of mystique because of the sense of their fundamental nature. Hence, there are plenty of books on the topics of cosmology and the ultimate constituents of matter, of the power of numbers and number theory, or on the provenance of humans, but there are fewer books on issues such as soft condensed matter, medical physics or atmospheric physics, supramolecular chemistry or microbiology. It is perhaps a brute fact (if there are any) that some areas of science are considered as "trendy" or "fundamental" where others are seen as more mundane. The choice of scientific topics ripe for popular science books usually is such that the topics covered are usually ones of long-settled science, or they are about questions verging on the philosophical and of little practical significance for everyday life. This is in stark contrast to the criticism that scientists often level at philosophers, that they are not practical enough.

As a result of this the public, with no grasp of the adversarial and contested nature of the production of scientific knowledge, is left with the impression that scientists usually speak with one voice, and that controversies in science are simply a matter of waiting for the cleverest person to come up with the optimal solution. Furthermore, any rivalries that do make it to popular culture are usually idealised accounts and do not give much insight on the modern culture of rival research groups and their inner workings. Science-in-the-making, that is science that is done through exchanges in scientific journals, conference presentations or laboratory basements is something very remote from the average lay-person, something which most lay-people are never confronted with. Which leads me into the second topic, that of the opaqueness of the scientific activity.

2.5.2 Opaqueness and inaccessibility of scientific knowledge production

The second related topic, introduced in the previous paragraph, is that of the opaqueness of science-in-the-making for the lay-person. Hidden away from the spotlight before presentation to a lay public, the everyday work of a scientist is totally inaccessible to the lay-person. In a way, even if the scientist wishes to describe their work to lay-people, they will often find themselves unable to, given that they are schooled in a different language and way of viewing the world. Scientific papers, even when making simple points and arguments, are directed at peers enmeshed in the specific language of the particular scientific discipline. The public does not access science in its primordial form, but rather through several levels of interpretation, after the involvement of many intermediates, such as science communicators and science journalists.

The inaccessibility of the language and the forms of communication that science uses to the lay-person makes it easier for the various deficits to become ossified. Given a few flashy names such as those of Isaac Newton and Charles Darwin, a brief no-miracles argument about the wonders of the technological world that surrounds us, and a few exotic sounding (usually) long words, the lay-person is left to consider themselves as somehow mentally deficient when in the same room with somebody who purports to be a scientist[21]. This communicative situation where scientists and lay-people effectively fail to understand each other despite their best intentions, often gets construed not as a problem of translation but as a lack of intellectual acumen on the part of the lay-people, something which leads directly to the various deficits discussed above. The situation is exacerbated when high-profile scientists make pronouncements such as "science is supposed to be difficult" (for example, see [Wolpert 2005; 1997], whilst at the same time other scientists (such as Sokal [2008], see above quotation) claim that science is nothing but clear thinking. These moves deprive the lay-person of any confidence with which to be able to confront scientists about any aspect of their work.[22] Hence, the deficit is almost self-imposed on the part of lay-people, it becomes a self-fulfilling prophecy.

In preparing the ground for a move back to a discussion of democracy in the next chapter, I will examine a final effect of the deficits discussed above, that of the denial of the possibility of constructive criticism on the part of lay-people.

2.5.3 Denying the possibility for criticism

As discussed above, one of the techniques found in the excerpts above is the bombardment of the keen lay reader of popular science books with a barrage of scientific facts, often in the same phrase. This is in stark contrast to scientific papers, which are often very specialised and in any case go to great lengths to support factual claims through references in the literature. However, as discussed in the preceding section, scientific papers are unavailable to lay readers either because of lack of access, or because of translation problems.

---

[21] Unfortunately, the same, if not a more pronounced effect, usually happens when one pronounces themselves as a "philosopher".

[22] Things actually get worse: because science is perceived as difficult and its practitioners as endowed with additional mental capacities, lay-people often resort to regard them as knowledgeable on matters outside their area of expertise, giving their opinion on any matter extra weight, sometimes unjustifiably so (especially given that an expert in one area is no different to a lay-person in an area outside that of their expertise).

Hence a real problem is created. The lay reader is faced with a plethora of impressive facts, but they have no way of corroborating them or of actually implementing the scientific method and deciding for themselves whether the facts or arguments presented are credible and truthful or not. Put simply, the lay reader often cannot distinguish between clear and muddled scientific thinking, given that they often lack the basics to understand even what the issues being debated are.

To this, it may be retorted that scientists are also being asked to take the word of their colleagues on trust, as exemplified by the use of scientific papers or of large collaborations in experiments such as those conducted at CERN (Hardwig 1991). However, as per the results of the survey studies that show that trust in scientists and science as an institution is not grounded on sound knowledge of the processes of scientific knowledge production, it can be claimed that there is a qualitative difference between scientists taking their colleagues work on trust and lay-people doing so. The difference is that scientists are more sophisticated in the way they receive their colleagues' testimony and they are more aware of the possibility of fraud, with Hardwig (1991, 702–703) noting that a significant amount of scientists knew of cases of falsification of data and of plagiarism.

Furthermore, there is a difference in degree as to the sources of information. The scientific paper is totally inaccessible to the lay-people, both in terms of language (as discussed above) and often of availability of the actual paper, whereas it is available in both senses to the scientist interested in verifying a certain claim. The division of intellectual labour, as discussed above, makes it so that the lay-person is simply asked to accept the finished product, often without the accompanying uncertainty, through the mediation of science popularisers or scientific journalists. In this sense the public simply lacks the resources to evaluate the scientific claim it is presented with. This situation is in some ways, despite professing to the opposite, similar to that of truth through revelation, in the sense that the lay-person is not evaluating any arguments for themselves or verifying any scientific facts.

This hapless situation is not helped even by accessible books written in an entertaining style, such as the excerpt given by Steven Jones above. To reiterate, it is most often the case that the lay-person cannot engage with the expert not because they do not have the intellectual ability to do so, but because they often cannot follow the argument because of the language that is being used.[23] An example of this is the widespread misconceptions

[23] To use a crude analogy, it may turn out that for a lay-person, attempting to

about probabilistic reasoning, especially on the topic of the chances of inheriting genetic diseases, and the comparison with many lay-people's shrewdness in calculating odds when betting on football scores or on horse-racing (on cultural differences as regards the link between probabilistic thinking and gambling, see for example [Lau and Ranyard 2005]).

Such a gap between experts and lay-people creates a democratic deficit, given that it severely restricts the possibility of substantial discussion of technical matters only among the class of the experts who are encultured (see for example Collins [1985] on the resolution of a scientific controversy) in the particular domain. From then on, for politicians wishing to make a decision on behalf of the whole population, it is a matter of appointing the experts judged as more appropriate to the issue at hand, hence reinforcing the bureaucracy and elitism described in Chapter One.

In the next chapter, I will move on from the discussion of the problem into a beginning of an answer, by promoting the twin themes of democracy and communitarian epistemology. I will link science and citizenship through argumentation, before I go on to identify school science education as the platform through which future citizens may obtain perceptions of science conducive to a more thoroughly democratic society. Furthermore, I will explore teaching of models within school science, something which will enable me both to develop some philosophical implications of communitarian epistemology, as well as the issue of intellectual safety in the absence of truth. This will pave the way for the defence of an epistemological view privileging the community in subsequent chapters.

---

follow a discussion by genetics experts may be akin more to attempting to follow a discussion on football conducted in a foreign language, rather than attempting to follow an exchange between Aristotle and Plato on the subject of the Forms.

CHAPTER THREE

SCIENCE, EDUCATION AND CITIZENSHIP

3.1 Introduction

At this point, I will pause and take stock of what has been achieved in the previous two chapters, in order to enable the reader to get a better grasp of my argument.

I began my argument in the first chapter by highlighting a social problem, that of the formation of scientific elites within societies otherwise labelled as "democratic". The existence of such elites comes to the forth especially in cases where they are called upon to assist policy-making. Having proposed that most current models of the policy/science interface rule out input from lay-people and identifying this conclusion as going against the ideals of participatory democracy, I concluded that a first step towards introducing the lay public into public policy-making would be to provide them with a sound understanding of the processes that lead into knowledge production in scientific settings.

Before answering the question of how this can be done in practice, I addressed the questions of actual transparency in the production of scientific knowledge and the grounding of trust in science as an institution and scientists as its practitioners. I used data from surveys conducted in the EU, US and UK and laid down their results as supporting the conclusion that there is indeed a lack of transparency built into the current models of science communication which are based on deficit models. I criticised science communication and public understanding of science efforts, before laying down some examples of instances of popular science writers who promote an elitist image of science. In this Chapter I will begin to explain how this problem can be countered. Hence, I will begin to consider in concrete terms the democratisation of scientific knowledge production and dissemination. After discussing some efforts in the democratisation of science for adults, I will identify the school as the locus where citizens can be "forged" who have an understanding of scientific knowledge and of participatory democracy as notions that reinforce each other. I will discuss briefly the philosophy of radical education, before moving onto linking science education with (democratic) citizenship education through their common insistence on argumentation as constitutive of scientific as well as democratic deliberative activities. Having established such a link, I will embark on a lengthy case-study on the topic of scientific models both in science education and in the philosophy of science. The case-study will serve a number of purposes: firstly, it will allow me to discuss a concrete problem in the teaching of science, that of the understanding and misunderstanding of scientific models by future lay-people; secondly, it will allow me to embark on a philosophical discussion on models and hence allow me to expose some of the metaphysical and other implications of communitarian epistemology, that is the philosophical thesis that I will present in the next chapters; thirdly, it will allow me to discuss an important implication that my view has on the topic of intellectual safety, a topic that will serve to link communitarian epistemology to democracy; finally, it will allow me to give a concrete proposal as to how the initial problem can be solved within the classroom.

3.2 Teaching science for citizenship

In this section I will attempt to link citizenship education to science education. I will begin by briefly describing some current efforts in enhancing active citizenship through debating over technoscientific issues pertinent to the livelihoods of all citizens, before briefly explaining why I choose school as the ideal locus for the forging of science-aware citizens. In the second sub-section I will elucidate the notion of argumentation being the common thread between science and citizenship.

3.2.1 Existing schemes using science for the enhancement of participatory democratic citizenship

The idea of science being more attentive and in close dialogue and interaction with society has been gaining ground in the last twenty years, the criticisms of Section 2.3.4 about the honesty of the whole enterprise notwithstanding. There have been various schemes designed to get citizens interested and competent enough to exercise their democratic right of being informed and being asked to deliberate on burning issues with a scientific dimension. Schemes such as Democ[1], used by the British

[1] Democ is a card game that consists, according to its originators (New Economics Foundation 2012), of information on a topic on a pack-of-cards format and of an

Council, up to science shops which bring together university research scientists and lay-people in constructive dialogue; as well as more comprehensive efforts such as the UK National Consensus Conference on Plant Biotechnology[2].

The topics that are discussed in these schemes are usually topics combining technoscience and societal problems, with the latter related to moral, aesthetic or practical issues that may form the subject of citizens' concerns. Examples of topics that have been discussed down the years are questions such as whether cloning should be allowed for medical reasons, whether the introduction of genetically modified organisms into the food chain should be allowed, the question of obesity as a societal problem, whether a particular country ought to be a pioneer in synthetic biology and so on.

A key pioneer in this area is the Danish Board of Technology, (Danish Board of Technology 2011a) which lists close to a dozen different methods designed to enhance participation in science-based policy-making and put in use over the last decade and a a half. I will sketch briefly three of these methods, before discerning some of their general characteristics which I see as positively contributing to the notion of science for the promotion of citizenship.

A very simple and effective method is the café seminar, which should ideally take place in a real café, as one of its chief aims is to generate informal dialogue in relaxed settings. Its originators stress that such a seminar is designed to raise awareness and enhance understanding of a technoscientific issue, rather than provide any sort of solution. The seminar itself can accommodate between 25 and 100 "guests" arranged in tables of five, with the guests taking turns in changing tables after each "session". Some reading material and questions are prepared by the organisers and experts in advance, however this material serves only to start the discussion on the topic. An example of a topic discussed in a café seminar is the vulnerability of IT systems to attack, which formed the subject of a café seminar held on the 7th October 2003 (Danish Board of Technology 2011b).

A more formal method is the Citizens' Hearing, which brings together politicians and lay-people in a two-way process of communication with the primary goal of building bridges between them. A further and more immediate goal is that of getting citizens to formulate their own suggestions towards solving a technoscientific-political problem and to

---

"interactive and engaged process to support deliberation in small groups".

[2] http://www.ncbe.reading.ac.uk/ncbe/gmfood/conference.html

make sure that politicians do actually hear these solutions. As part of the design of the hearing, the outcome is a product solely of dialogue between citizens, without the interference of experts or politicians. The planning group that organises the hearing consists of experts working in the field, however their sole task is to choose the topic of the discussion and to formulate a discussion paper, which is sent to participants before the meeting. In order to make sure that participation is as representative of society as possible, interested participants are chosen by random invitation as well as by newspaper advertisements (Danish Board of Technology 2011c). Examples of topics tackled through this method include alternative medicine, education and surveillance and privacy.

A final method which has become more widespread is the Consensus Conference. According to the Danish Board of Technology, at a Consensus Conference

> Citizens contribute by making their views known in the form of visions, concerns, values, holistic appraisal and everyday experiences. The consensus conference method is based on the premise that technological assessment cannot be limited to the legislative domain. (Danish Board of Technology 2011d)

During the lead-up to the conference a panel of randomly selected lay-people is given the chance, through two preparatory sessions to gain knowledge on the topic from technical experts in the field, and then they are asked to formulate a set of questions to be presented to the expert panel. According to the Board,

> These must be questions they seek answers to in order to assess a given topic and in order to make their recommendations to politicians regarding future control and development in the area. Different kinds of experts with opposing views are summoned to answer the panel's questions. (Danish Board of Technology 2011d)

The conference itself takes place during three days, from Saturday till Monday, with up to 25 experts answering the panel's questions on Saturday, whilst on Sunday the panel withdraws in order to discuss the answers given by the experts and draft a final document which will be presented back to the experts for the corrections of technical mistakes and to the audience for discussion on Monday. The final session on Monday is open to the public, politicians and the media, and a debate ensures that the lay citizen panel has the chance to defend its final recommendations.

A general point also concerns the aims of these processes of active public engagement in what often tend to be technical debates, or in any case the bringing together of technical issues and societal problems. As regards what is achieved through these processes, the result is often a recommendation that is presented as representative of the people's will to a public authority, be it a national government or a local council. The conclusions and recommendations of these exercises are, more often than not, not binding to the authorities, however they are taken seriously by members of the political class. Examples of issues tackled through Consensus Conferences include gene therapy and genetically modified organisms in the food chain.

The above schemes are exemplary of how citizens may exercise their democratic right of taking active part in democratic deliberation over issues incorporating a technoscientific element. However, the real question is whether citizens can be systematically encultured into openly discussing and deliberating on social issues with a technoscientific element, and, taking into account the criticisms at the "Science and Society" agenda in Chapter Two, whether they will demand that their political representatives listen to such deliberations.

The ideal locus where a systematic effort at such an undertaking can be taken on board seems to me to be through the two cycles of primary and secondary mandatory education. Viewed through a slightly idealistic lens, part of the mission of school education would naturally be to produce model citizens, a realization that is substantiated more with the introduction, in many countries, of citizenship education. Furthermore, school education, being mandatory, has the added benefit that it actually does reach all future citizens at an age when their social and cultural identity is still being forged. Finally, school education is also the forum where future citizens have their first systematic contact with the natural sciences, in the form of science education.

I therefore will now turn my attention to primary and secondary school education. In the following subsection I will discuss a certain tradition in educational philosophy that adheres to the ideals of participatory democracy, before moving on to the discussion of science education as inextricably linked to citizen education.

3.2.2 Radical education

The origins of the radical tradition in education are to be found in the writings of John Dewey on the topic of democracy and education[3]. Dewey mounted a spirited critique of the elitist conceptions of democracy that

[3] See especially his "Democracy and Education" (Dewey 2004)

were dominant in his time and advocated a return to participatory democracy through the viewing of democracy as a way of living. According to one commentator (Childs 1989, 443), Dewey realised that the conditions current in American society were detrimental to the American democratic heritage, and that the dominant social classes were seeking to limit free expression of thought. To this Dewey (1981, 446) countered that schooling should be viewed as a form of community life in which students

> learn to understand themselves as democratic individuals by becoming members of a community in which the problems of communal life are resolved through collective deliberation and a shared concern for the common good. (cited in Carr and Hartnett 1996, 63)

The locus where students are encultured into the democratic praxis is the common school, where students from all social backgrounds freely mix and create a community ripe for practising participatory democratic ideals. Both the locus, the common school, and the content, that of practising democratic ideals, come into direct contrast with existing accounts of both democracy and schooling, hence the label of "radical", even though the definition of radical education is widely contested (Fielding 2008, 543–544).

An issue that arises within such a debate is whether schools ought to trail social change and prepare students for their role as future citizens within the existing social order or whether schooling ought to prefigure, or act out the desired future change. In this, traditional schooling opts for a more vocational schooling that prepares and moulds students into future consumers and workers, as opposed to radical education which has as its aim the production of democratic citizens in anticipation of social change towards a more participatory democratic society.

The relationship between radical education and science education is one that hasn't been explored in depth, however some elements of prefigurative practice that were practised at the St. George in-the East Secondary School in Stepney, London between 1945 and 1955 may be seen to affect science education as well. According to Fielding (2008, 551) the curriculum was chosen by the students themselves and knowledge was co-produced and co-constructed, and furthermore a strong sense of collegiality and community replaced fear of failure and punishment as well as a sense of endless competition.

To be sure, in his educational approach Dewey advocated what Childs (1989, 421–422) called "the method of experience" which, with its deeply embedded empiricism and its contrast to supernaturalism may be seen as quite germane to what is often labelled (but rejected ultimately in the presently) as "the scientific method". However, Dewey also considered the individual mind as firmly rooted within its natural and social environment, and was totally against authoritarianism and rote learning in the classroom, something which ensures that the teaching of the natural sciences would be in line with my advocacy of science as conducive to a healthy democratic citizenship. I will nonetheless in the next section address some worries about science and citizenship not entirely being congruent with each other.

3.2.3 Science education and citizenship: uneasy bedfellows?

The relationship between science and citizenship is quite a complex one, fraught with political and other dangers. As has been shown in Chapter Two, the sheer individual brilliance of prominent past scientists, both as solitary scientific geniuses as well as people who benefited the whole of society are often exaggerated. Furthermore it is often implicitly claimed, at least at the level of lay-people, that a solution to a societal problem proposed by scientific experts is a priori a better one than one arrived through by other means. These perceptions of science have been explored and criticised in the previous chapter. The expectations that lay-people have of science presents a further criticism; they often expect scientists to give answers to what are essentially trans-scientific or even philosophical questions, that is questions that cannot be answered by scientific means alone or are even inappropriate as topics amenable to a scientific analysis.

Indeed, a more interactive engagement with science and an exposure to the uncertainties inherent in science may lead to a general scepticism, which leads one commentator to worry that:

> Any such disillusionment stemming from immersion in public controversies might have intriguing implications if it were replicated in the classroom... if healthy scepticism became transformed into intellectual cynicism and disbelief amongst students, most teachers of science would see (surely with justification) the resultant marginalisation or rejection of science and scientific expertise as unhelpful. ([Thomas 2000, 142] cited in [Ryder 2002, 159])

Hence it seems that science education is caught between two extremes: the first, idealistic (or scientistic) perception of science would claim at its most extreme that there are no trans-scientific issues; that science done properly can give us the optimal answers to most important questions of social importance. Such an extreme promotes the elitist perception of science as an institution (Feyerabend 1978). The second extreme is that which, according to the more interactive model regarding the relationship between science and society, gives emphasis to current controversial scientific topics, with the danger of producing citizens wholly disillusioned with science. Such a citizenry would be dangerous in that it will be indifferent to science without engaging with it, that is without giving it a chance to prove its worth[4]. This second extreme becomes even more pernicious when one considers that such a cynicism could go beyond a healthy level and may lead into a situation in the future where scientific institutions will struggle to continue research generally admitted as worthwhile for humanity for lack of professional researchers; or perhaps more alarmingly people would die from illnesses such as measles because parents refuse to vaccinate their children against such diseases, by failing to comply with the best scientific advice from the field of biomedicine.

A middle way of utilizing science education therefore has to be introduced, one that would avoid the pitfalls of the two extremes laid down above. One way would be to think of scientific knowledge as part of a coherent whole that would be labelled "citizen thinking". Jenkins, in describing the place of scientific knowledge in such a body of knowledge asserts that,

> During the course of their personal, working and social lives, all citizens construct a body of practical knowledge, tested and validated against their individual and collective experience. In deciding how and when to act in practical matters that have a scientific dimension, scientific knowledge is considered alongside this other, experiential and personal knowledge base.(Jenkins 1999, 15)

The idea would be that scientific knowledge will be collected only when it is required for action, and that, furthermore, such scientific knowledge may be distorted and tailored to fit in with the personal stock of practical knowledge of the person in question. Jenkins gives examples where rejected scientific theories or utterly unscientific claims are adopted by lay-people and used as a basis for lay theories of things such as electricity or heat. These theories, based on notions such as the "dissipation" of electricity into the ground or a "flow" theory of heat are made consistent with the specified people's practical knowledge and lead to sound and prudent decision-making, even though their basic assumptions have long ago ceased to form part of the body of scientific

---

[4] See the "Indifferent" segment of the UK population in the UK PAS 2011 report.

knowledge.

A criticism of such a view would be that science education would turn out to be irrelevant, in the sense that the stock of practical knowledge would develop mostly when the person in question leaves school, or in any case the stock of practical knowledge will co-evolve with the experiences of the person. Such a development would not be catastrophic, as the polemics of Ivan Illich (1971) or, more recently, the work of Nuñes et al. (Nunes, Schliemann, and Carraher 1993) on street mathematics has shown, however, in the context of the present work, I will support a position which takes school education for granted and attempts to provide a positive alternative that makes use of it. A reason for this is that it is entirely possible that the people described in the studies of Illich and Nuñes are often disenfranchised and powerless as citizens, in the sense that their opinions are often totally ignored by those in power. As opposed to this, my thesis is intended to produce an alternative that would empower all citizens in the context of a 21st century fully technoscientifically-laden participatory democratic society.

3.2.4 Argumentation as necessary for both science and citizenship

A thread implicit in my attempt to sketch scientific input into state policy as a threat to democracy, stems from the notion that argumentation in which all stakeholders (especially lay ones) will be present and equal is what lies at the heart of democracy, even in its modern sophisticated reincarnations. Argumentation both in formal and informal settings seems to be a thread running through the schemes designed to enhance democratic citizenship among adults described above.

The promotion of argumentation contained in the subject of science education is already a burgeoning research programme in science education (see for example Osborne et al (2004)). At this point it is worth clarifying that by argumentation I do mean a cooperative social activity designed to ideally lead to consensus, as opposed to "having an argument" as is used colloquially to designate an activity akin to a shouting match. Hence the notion of the argumentation process as a collaborative one with the aim of reaching a consensual conclusion or simply agree to disagree.

Argumentation is thus a social activity in the sense that it differs to solitary internalised thought processes, and from a cognitive perspective,

> to the extent that argument involves the public exercise of reasoning (Kuhn 1992; Billig 1987), lessons involving argument will require children to externalise their thinking. Such externalisation requires a move from the

> *intra*-psychological plane, and rhetorical argument, to the *inter*-psychological and dialogic argument (Vygotskiĭ 1978). (Erduran, Simon, and Osborne 2004, 3)

Hence, according to the above quotation, the process of argumentation lies at the essence of communication, as it brings thoughts out into the world, and is the only way through which concerted social action can come about.

However, the structure analysis of argumentation and whether there are any stable rules in it, such as logical rules or other "rules of communication" is a much more woolly issue. Even though a strong current in argumentation within the science education literature focusses on Toulmin's Argument Pattern (TAP) (Toulmin 1958), which uses concepts such as "Claim", "Warrant", "Data", "Rebuttal" and "Backing", some of the main authors themselves acknowledge that models building on these concepts are mere simplifications for the sake of analysis, and that argumentation in real situations is much more complex, not least because it is not clear even when using the above claims to discern what constitutes a claim or a warrant, and so on.

Erduran et al. (2004, 3) further remark that schemes other than Toulmin's exist that provide heuristics for the analysis of argumentation, such as Latour's (1987) scheme for scientific argumentation or Walton's scheme on presumptive reasoning (1996).

Given my affinity to relativist theses, which will be unfurled in the following chapters, it is perhaps no surprise that I will take a relativist position on the topic of argumentation. Therefore, my position is that the rules of argumentation are relative to different communities of people or relative to different time-slices of the same community. I do claim that the capacities for argumentation are probably very easy to learn, in the sense that the process of acculturation into the rules of argumentation of a given community probably does not require any skills and is within the grasp of most, if not all standard humans of all ages. Nevertheless this does not in any way imply that we humans carry with us an innate capacity for argumentation. Rather, my claim is that the capacity for argumentation skills is developed as soon as babies enter into situations where they learn to externalise or communicate their thoughts or desires, such as when they need to give the message to their parents that they're hungry or in pain. Of course, what is developed then by infants is the capacity to look for and learn the 'rules of engagement' of different communities and different situations, rather than knowledge of these rules themselves. To give an example, a situation when children are deciding and carrying out a game of football is different to a case where two people are discussing politics at a café which is in turn different from philosophers discussing the metaphysics of causation during a research seminar at the University of Bristol.

If, however, argumentation is so ubiquitous then what is special about science and citizenship in relation to it? Well, the problem is not so much that argumentation is present and constitutive both of science and of democratic citizenship, but that it be recognised as having such an important role. My claim is that doing science or practising democratic citizenship ought to be seen and analysed primarily as engaging in forms of public argumentation, perhaps over a limited domain which is either framed by the focus on empirical evidence, as is the case in science, or over action in the public sphere for the sake of common interests, as is the case for citizenship.

This is in opposition to the current understanding of both. On the one hand, science is regarded as primarily an activity oriented towards uncovering regularities in the natural world. On the other hand, citizenship and civic engagement is too often viewed in terms of the individual or at best the individual's engagement with their community, as in the psychological concept of civic engagement (Verba, Schlozman, and Brady 1995) or the perception of citizenship education's aims as to create autonomous individuals capable of pursuing their own political interests in the public sphere (Nie, Junn, and Stehlik-Barry 1996).

The difference between these two perceptions, to reiterate, is a shift of focus from the individual subject and their actions, either in total isolation from the community or at best in an interaction with it with sharp borders between the personal and the public sphere, to a focus on the community as the primary object and the individual viewed only as a part of the community.

In that respect, the door opens for my case-study-cum-proposal for school science education, with the proposal being that learning science ought to provide school children with a test-case for public argumentation, by focussing on argumentation based on empirical evidence. This it can achieve through highlighting the social aspects which are constitutive of science[5] and by providing topics for conversation and the co-creation of knowledge based on consensus. Through the following exposition on the topic of models in science education, as mentioned above, I have several key aims, such as to discuss a concrete problem and provide a solution to it congruent to my avowed defence of communitarian epistemology as conducive to a more democratic polity, to sketch a communitarian position on the topic of the ontology of models, and finally to discuss a

[5] I will return to this topic in the next chapter.

fundamental implication of communitarian epistemology, that of teaching without an emphasis on truth but rather on consensus arrived at through democratic means.

3.3 A case-study in socially-minded science education

3.3.1 Introduction: Water and models

This case-study concerns a thorny topic within the philosophy of science, that of models, and has among its aims that of giving a communitarian epistemology-flavoured proposal for the teaching of the use of models in science. I will take the model of the water molecule as an example which will lead me into the study of both the philosophic and the educational aspects of models. It should be noted at the outset, however, that the water molecule model is a special case of a model as used in science, as it is what Contessa (2010, 217) labels a material model, that is a model only consisting of an entity, rather than a fictional model, which consists both of a model entity and of a theoretical description and/ or mathematical equations. The distinction between model entities and fictional models, with the former being parts of the latter, should be kept in mind in the following discussion of the ontology of models.

In my philosophical discussion about models I will also use other models such as the linear harmonic oscillator, as well as models of the atom. These models are fictional models, (once again, following Contessa's classification – for the time being I do not wish to commit myself to any metaphysical view of models, something which I will do in following sections) that is they involve theoretical descriptions as well as the model entities, however for purposes of illustration I will for the moment focus on the water molecule. After discussing the ontology of models I will return to my positive proposal for using the topic of models to incorporate communitarian aspects into the teaching of chemistry.

Maybe the most recognizable formula in the whole of chemistry is the chemical formula of water, H_2O. That much is clear. What is not clear, however, is what this formula actually tells us. I will focus on the microscopic interpretation of the formula, which is roughly along the lines that there is such a thing as a water molecule which consists of two atoms of hydrogen and of one atom of oxygen, in an arrangement akin to Illustration (1).

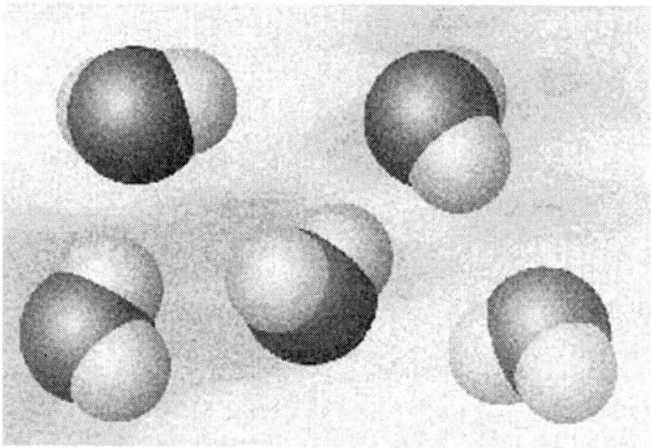

Illustration 1: "Image" of water molecules

This description is at least inaccurate, if not plainly false. Needham (2000, 19), citing van Brakel (1986, 299), argues that in a given sample of pure water there will be molecular structures such as hydrogen-bonded polymers made up of various numbers of the molecular monomer, as well as ions such as positively charged hydrogen ions and negatively charged hydroxyl ions, as well as their polymeric structures. Furthermore, these molecular structures are not stable in time, but change as the molecules move around and associate or dissociate. In short, Image 1 is not a scaled-up "picture" of water molecules at any given time, and H_2O is not an accurate representation of water at the molecular level[6]. Figure (1) is hence not something we would see if we magnified a water sample billions of times, but is a rather simplified model of water, appropriate for certain purposes and inappropriate for others. Or, to put it more formally and with an allusion to some early work in the philosophy of models, Figure (1) simply presents an example of an analogy within science. Mary Hesse, following Max Black's work on metaphors, introduced the view of scientific models as analogies and metaphors that play an indispensable explanatory function in the sciences (Hesse 1966). Hesse offers a defence of analogical reasoning, however I will not expand on this topic in the presently, instead focussing on perspectives from within the science education literature on the topic of models.

[6] What, of course, remains true, is another interpretation of the formula H_2O as standing for the statement that water by volume of reacting gases consists of two parts of hydrogen for one part of oxygen.

The problem that I will thus concern myself with is the mistaken perception by lay-people that Figure (1) is indeed a scaled-up picture of water molecules, and whether an approach in science education can provide the lay-person with a better understanding of the use of models in science. I will now first describe the problem, then delve into the philosophical issues associated with models before laying down a proposal for enabling students to be more competent in their use of models through practices in the classroom.

3.3.2 The educational problem with models

I have used the above short study of the water molecule model to flag up a problem concerning the use of models in science education. The problem is the literal interpretation of models; that is the view that models are essentially scale models of reality. Grosslight et al. (1991, 818-821) have identified three levels of understanding of models for educational purposes, and in their study found that a significant proportion of students do not move beyond the first or "mixed" first and second levels of understanding. The three levels identified by Grosslight et al. are the following:

Level 1: models are thought of as existing concrete objects which are copies of reality

Level 2: the reality modelled is still the main motivation for the construction of model, however the purpose and ideas of the modeller begin to play a role. The models are still primarily thought of as existing concrete models which are representations of reality, even if some aspects may be intentionally abstracted away or idealised

Level 3: the ideas and purposes of the modeller are the main motivation for the construction of the model, rather than the faithfulness to the model's target system. Several different models of the same target system can be constructed according to the needs and purposes of the modeller, and these models can be tested and manipulated on their own merits.

Grosslight et al. see the three levels in a hierarchical relationship from level 1 to level 3, and remark that it is mostly expert scientists regularly constructing and manipulating models that reach the third level of understanding of models, whereas most students' understanding of models is to be found at the first level or somewhere between the first and the second level. From this, and from the fact that most lay-people do not study science after secondary school, it follows that lay-peoples' and scientists' perception of scientific models is different.

This misconception of models infects lay-people's understanding of science and sometimes creates high expectations of certainty from scientists. These expectations arise because of the failure to appreciate that reality outside the lab is much more complicated than can be presently studied, and the best that scientists can do is base their predictions on simplified models that make the problems mathematically tractable.

An example of the limits of scientific modelling is the current situation in climate change theory and of projections of the effects of global warming in the future. Because the target system is very complex, current studies and simulations are limited by computational power, and every new revision or addition to the current models may bring radically different predictions. The effect of public announcements of diverging results without the necessary stress on their contextual and model-dependent nature is the erosion of trust in scientists' announcements among lay-people, who only become aware of the predictions without realising the inherent limitations of the modelling process and the uncertainty involved. Hence, I conclude that a better understanding of scientific modelling is needed in order to avoid misconceptions about how science works and about the products of scientists. In the next section I will shine some light on discussions in the philosophy of science on the ontology of models, before moving on to look at the problem from the angle of science education, in accord with the discussion of the previous sections.

3.3.3 The philosophy of scientific models

Scientific modelling and the manipulation of scientific models take up a great deal of time and effort of practising scientists. According to one commentator, (Contessa 2010, 217-218) there are three different sorts of models employed within science; material models, such as the balls-and-sticks model of DNA or of water molecules; mathematical models, which consist of one or a set of equations, and finally fictional models, such as the ideal pendulum or the Bohr model of the atom. These models may consist both of a model entity (for example the ideal pendulum in the linear harmonic oscillator model) and of equations or theoretical descriptions (for example the equations of motion for the same example). The latter consist of most models actually used within science yet their ontological status is most problematic. In the recent discussions on the ontology of models it is the model entities which are considered to be fictional, mostly by anti-realists about scientific models and about science and scientific entities in general.

The question of the ontology of scientific models is more pressing if one considers the mainstream semantic view of scientific theories, which claims that theories simply consist of the sets of models that render them true. This question has surprisingly attracted relatively little attention until recently, when two main schools of thought have emerged, along the realism vs anti-realism debate. The dominant view, mostly associated with the realist viewpoint, is that models are abstract objects (Giere 1990) whereas the other, anti-realist view that has been advanced lately is that models are fictional objects (Frigg 2010) being in a family resemblance relation to the descriptions and objects found in literary novels. What is sought for is the representation relation that models have vis-a-vis reality, given that they have, at least when used in scientific practice, a high explanatory power.

I will expand on a realist and an anti-realist account of the ontology of models, before turning on to briefly sketching my own stance on the issue.

3.3.3.1 Realism about models: models as abstract entities

The realist about scientific theories believes that scientific theories are literally true and the entities that they posit or trade in literally exist. Given that scientific theories have models as essential parts, (or, according to the semantic view, theories just are collections of abstract mathematical models) then the realist has to explain how it can be that scientific theories essentially involve entities which do not exist, at least not in the same way that electrons and atoms are said to exist. Scientific models such as the ideal pendulum, which consists of a point mass suspended from a massless string with no friction are nowhere to be found in the world as we know it. The problem here is that the basic characteristics that define the model (for example the pendulum), and at the same time the ones that give it its explanatory power, are characteristics that no body in the universe may possibly possess. As opposed to electrons which are essentially described by their possession of negative charge, a property which is clearly defined, accepted and regarded as existing, we know by design that no pendulum's bob is a point mass simply because in the real world there is no such thing as a point mass. Furthermore, it is arguable that the point mass is something which couldn't possibly exist, much like a round square could not possibly exist.

In the face of this, the realist move is to claim that models are much like abstract objects, such as numbers, and as such do not inhabit our own universe but "live" in a universe of abstract objects, which also contains entities such as numbers and geometrical points. In what follows in this section, I will focus on one argument provided by Psillos (2011), who argues that scientific realists can happily accept models qua real but abstract objects.

Psillos proposes that, given the central role that scientific models play within scientific practice, the scientific realist is faced with what he labels as the "Central Dilemma", which is the following: either accept that theories include things that are known not to exist (models), hence they are born false, or theories should not be read literally, hence scientific realism goes (2011, 3). The strategy that Psillos follows in order to get out of this dilemma is the following: first, admit that there is an abstract but real object (the model) that satisfies the description that emerges out of the two processes of model-building, (abstraction, in which some aspects of the system under study are omitted on purpose, and idealisation, in which some aspects of the system are wilfully misrepresented); then driving a wedge between the process of model-building, which, because of idealisation and abstraction favours seeing models as fictions, and the model as the product, which is best seen as an abstract but real object, standing in a representation relation to a worldly object. For example, according to Psillos, the simple harmonic oscillator **A**, whose description **D** is produced by considering a real pendulum (object **B**) and stripping some properties (e.g. colour) and wilfully distorting others (considering the bob as a point mass) is an abstract object considered to stand in an appropriate representation relation with the real pendulum **B** that gave birth to it. The description **D** is such that it is not an exact and accurate description of the concrete object **B**, however, it is true of the real but abstract object **A.** According to Psillos, the move of admitting models as abstract objects comes with the price of introducing an extra layer of mediation in the representation of the real world through scientific theorising, however, this price is worth paying.

A more serious price Psillos seems ready to pay in order to maintain his scientific realism about models is the Quinean (see for example [Quine 1960, 22]) admission that there is no theory-free standpoint from which one may be able to decide what entities are real. This move is intended to ease pressure on the realist and to help pave a way out of the central dilemma described above. However, the admission that there is no theory-free standpoint from which to judge what exists and what does not is a standard relativist assertion, and as such goes directly against, or at least severely weakens the scientific realist line and brings it closer to a strong relativist one. As I perceive it, ontological commitment and truth go hand in hand, and it is not the case that one may claim theory-ladenness for ontological commitment and not admit it as regards truth. Hence I claim that the specific scientific realist argument about models ends up not differing much from the relativist line on the same topic, and this I regard as bad news for the realist. I will now move on to a second attempt to give an explanation of what models may be, this time following the anti-realist view of models as fictions.

3.3.3.2 An anti-realist account of models: models as fictions

At this point, it is worth restating the problematic that set us off. What is sought is a "user-friendly" account of models that will be fit and fruitful for teaching science, whilst at the same time being faithful to scientific practice. It seems that a fiction view of scientific models can capture both aspects of this condition.

The philosophical account which I will put forward is one which regards models in the same ontological status as entities occurring in literary fiction works, using the resources of pretence theory (Walton 1990). As regards their ontology, models are viewed as hypothetical physical systems, which are just like concrete physical systems apart from the fact that they lack the property of existence. This is intended to capture the intuition that we can learn a lot from a model about target laboratory systems, for example we can learn a lot about pendula in laboratories by studying the harmonic oscillator. This is in analogy to the characters in works of fiction, which have the properties of real people, apart from the fact that they don't exist in the actual world.

According to Frigg (2010a, 257-258), who advances the account that I am elaborating here, there are three main motivations for treating models as fictions. First, there is nothing in the real world corresponding directly with the model, and this is known to the people who manipulate the model. There is no such thing as a perfectly spherical planet or an infinite potential well, and the scientists using these models are perfectly aware of this. Secondly, there is more to the model than described in its description, and one can learn about this extra content by following explicit or not so explicit rules of inference. Such rules are to be drawn from the content of the theory, for example when describing the two-body model of the Earth and the moon, the resources of Newtonian mechanics are drawn in to make inferences about the model. Finally, good models tell us things about the real world, just like good works of fiction do. This is achieved both straightforwardly by the model description, which contains many facts, as well as more implicitly by the deductions made by the model. In analogy with works of fiction, Dickens intends the reader to learn about conditions of life in 19th century London and adds quite many facts as to this effect, in the same way that Niels Bohr embeds the relevant knowledge of facts about atoms in his model of the atom. However, the presence of facts is not the only way that a person learns from the model qua work of fiction, as a lot of the truths that emerge from the model are derived by the application of rules of inference and by manipulation of the model.

The central notions of Walton's (1990) pretence theory, on which the account is based, are those of props and of principles of generation. A prop is any object that may affect our senses and thus prompt our imagination. An object becomes a prop if and only if a principle of generation is imposed upon it. This principle of generation prescribes what is to be imagined as a function of the prop, which is the object present. A person engaged in such an activity is thus engaged in a game of make-believe, and they are pretending to be in the world prescribed by the principle of generation. The principle of generation may be ad hoc, that is arbitrarily imposed by a small group and only valid for that group for a given amount of time. In such a case the game, according to Walton, is unauthorized. Examples of unauthorized games are children's games. To give an example, some children may decide to play with toy soldiers, and agree that a table-top stands for a sea in which books serve as islands, and that if the plastic figurines standing in for soldiers fall from the book tops then they drown and die. In this example, the table-top, books and toy soldiers all are props. The principles of generation is the prescription to imagine that the plastic soldiers stand in for real ones, that the books are to stand in for islands and that the table-top is to stand in for the sea. The game is unauthorized as most of the principles of generation are ad hoc, apart from the one about the toy-soldiers, something which I will return to shortly.

A prop, by definition, is a representation if it is part of an authorised game. Hence, the character Hercule Poirot is a representation, since the principles of generation regarding this fictional character are accessible to anybody fluent in English and furthermore, they have been stable since the first appearance of the character in the Agatha Christie novels. In the above-mentioned toy-soldier example the toy soldiers themselves are representations, since the fact that they are to be taken as standing in for real soldiers is obvious to everybody. In Walton's theory, the term "representation" does not signal a relationship between two things, but rather representations are simply things that possess the function of being props in authorised games of make-believe.

According to Frigg, (2010a, 259) truths in the model (fictional truths) are created by virtue of principles of generation and their actions upon props. There are two kinds of generated truths, directly generated truths and indirectly generated truths. Directly generated truths (called "primary" truths) follow immediately from the props and are associated with "principles of direct generation". Indirectly generated truths or implied truths follow from "principles of indirect generation" along with the application of rules of inference. To give an example from a work of fiction, it is a direct truth that a flower grew in the Little Prince's planet, however it is an indirect truth that the flower was greatly saddened and cried a lot when he left.

In the context of a physical model, such as Bohr's planetary motion model of the atom, the "principles of direct generation" would be the linguistic and mathematical conventions that allow us to understand the description of the model, whereas the "indirect principles of generation" would be the physical laws which allow us to infer further truths from the model. In another example that Frigg gives, of the familiar Newtonian two-body system, the verbal description of the model is the set of the primary truths, with the derived truths being the truths, such as the elliptical orbit, being generated through the "direct principles of generation" which are the linguistic conventions that allow understanding of the description, and the "indirect principles of generation" being Newton's Laws.

3.3.3.3 Frigg's schematic account of scientific modelling

Armed with the above theory, Frigg gives the following schematic account of scientific modelling (Figure 2, reproduced from [Frigg 2010a, 266]).

Frigg claims that any element of the above schema, apart from the target system, may legitimately be called "the model". It is worth remarking that the mathematical model structure is part of that and that furthermore two notions of representation are at play. P-Representation is representation as defined in Walton's theory, that is representation through props, whereas T-Representation is representation in the mainstream usage of the term within philosophy, and serves as the relation between the model system and the target system. Frigg uses an analogy with maps in order to explain the exact nature of T-representation, setting down the following condition for some X to T-represent a Y:

X T-represents Y iff
X denotes Y
and X comes with a key K that specifies how facts about X are to be translated into claims about Y

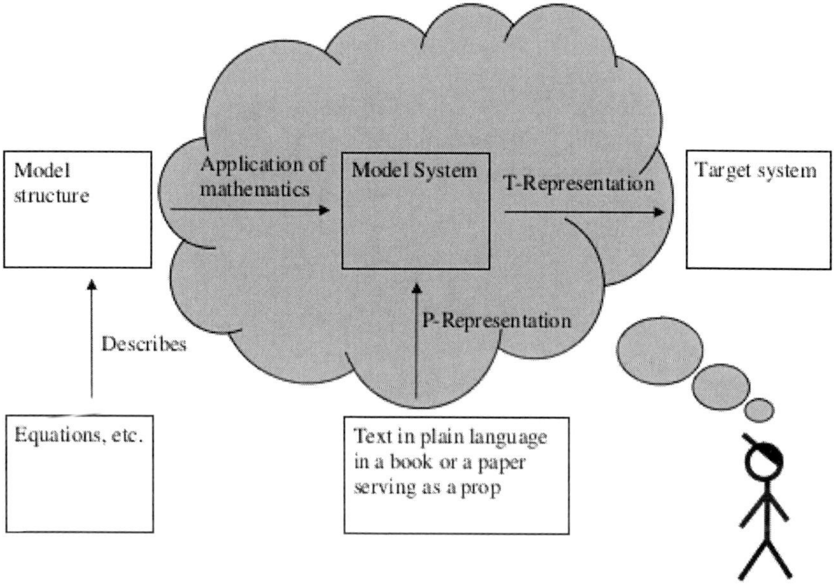

Illustration 2: Frigg's schema of the elem*ents of scientific modelling*. With kind permission from Springer Science and Business Media

The key is something like the conventions laid down in the key of a map, however Frigg notes that in scientific representation this key is often missing and is the subject of stipulation among scientists. This results in disagreement among scientists as to how to interpret the model, that is relate it to its target system. Frigg notes that T-representation is not a straightforward notion, and no general account can be given of it, but rather each case ought to be treated separately.

Frigg (2010b, 132–135) gives the example of the Earth-Sun system as an example of his account. I will follow his description in order to briefly present this example here. According to this example then, we begin by assuming that the only forces acting on the two bodies are the gravitational forces of each other, neglecting all other forces. We then assume that the two planets are perfect spheres with their masses equally distributed, something which allows us to see the two planets as point masses. We assume Newtonian mechanics in order to relate the force and the acceleration of the two bodies. The above description serves as a prop and we thus engage in an authorised game of make-believe. The direct rules of generation are simply the rules of English, and what we are asked to imagine is the model-system. The mathematical model-structure consists simply in the Newtonian equation, which applies to the model-system and is true of it. It is a formal principle of indirect generation, since it gives us the implied truth of the Earth's elliptical orbit around the sun. What remains to be specified in this example is the T-representation relation between the model and the system. The first step towards achieving this is done through the use of ordinary language in the key, through which we state that the body with mass m_e in the model corresponds to the actual planet Earth. The next step is to specify the abstractions and idealisations made for the sake of coming up with the model system, and to explain how these are to be understood. It is worth remarking, as Frigg does, the hypothetical character of the key, whilst at the same time stressing that it is not entirely arbitrary, but rather draws on background knowledge and is also validated with available data.

That the "models as fictions" account is anti-realist in nature comes from the fact that Frigg asks us not to assert truths within the model but to imagine them. According to Walton, such "truths in fiction" are not the same as everyday truth, in fact they are not even to be considered as species of truth. As regards gaining information about the target system from the model system and the associated metaphysical problem, Frigg claims that the problem goes away if we think of the information gained not in terms of being drawn from a comparison between two objects, but between two different sets of properties.

A final problem that comes associated with the models qua fictions account is that of the subjectivity of our imagination, that is how a single model might emerge out of the modelling process, rather than as many as there are modellers. The solution to this problem may be drawn from Walton's pretence theory and his notion of authorised games. Hence, in science, such a problem does not arise as the activity of scientific modelling is a rule-governed activity and the props and principles of generation are public, and as such they constrain what is to be imagined, at least in its scientifically relevant properties.

3.3.3.4 Realism, anti-realism and the ontology of models

In discussing the ontology of models, I have presented a realist account, that of models as abstract entities, and an anti-realist one, that of models as fictions. I will now attempt to sketch the communitarian epistemologist's position on the issue of the ontology of models, as this is the position that I will defend more explicitly in later chapters.

First of all, it is worth remarking that the communitarian epistemologist position comes devoid of any ontological commitments on models apart from the one that there are communities of enquirers that do not have direct epistemological access to reality. By this I mean that for scientists involved in scientific modelling, the ultimate arbiter as regards the relationship of their models to such a thing as "reality" is nothing other than consensus among their peers as to this effect. "Truth" and "correspondence to reality" are constructed by community consensus together based partly on the rules of inference of that particular group, as well as partly on the differences brought into the table by each of the participants' individual thought-processes and experiences. By this I want to bring forth one aspect of communitarian epistemology, in which it is opposed to certain varieties of anti-realism. This is that, whereas in certain varieties of anti-realism, there is "right" and "wrong" within a given conceptual scheme, in communitarian epistemology there is no such thing, but rather actors' interests and indeterminacy rule supreme, and the arbiter for "right" and "wrong" is the community of enquirers[7].

Given that the only ontological commitment of communitarian epistemology is to communities of enquirers, it follows that commitment to mind-independent abstract objects is blocked. However, Psillos' weakly realist position (weak as it admits dependence of the existence of theoretical posits on theoretical viewpoints) is naturally attractive to the communitarian epistemologist who sees a parallel with their own relativist line that there is no perception of reality independently of conceptual schemes.

On the other hand, the eschewal of any ontological commitments that comes along with Frigg's anti-realist account is also congenial to the communitarian epistemologist, who is attracted to Walton's pretence theory, pending of course on the assumption that "right" and "wrong", or "truth" and "falsehood" within games are subject to the community's consensus rather than derived directly by the principles of generation. This constructivist position, which places community consensus at its centre, comes into conflicts with some varieties of anti-realism, such as Putnam's internal realism. According to internal realism,

> Once we make clear how we are using "object" (or "exist"), the question "How many objects exist?" has an answer that is not at all a matter of "convention". That is why I say that this sort of example does not support cultural relativism. Of course, our concepts are culturally relative; but it does not follow that the truth or falsity of what we say using those concepts is simply "determined" by the culture. (Putnam 1987, 71)

---

[7] On this, I will have more to say in Chapter Five, where I highlight the strong community thesis in normativity.

This is exactly the sort of anti-realist feature that the communitarian epistemologist rejects, opting for convention all the way down.

As regards the question of the ontology of models then, the communitarian epistemologist can simply point out the diversity of uses into which they are put by different scientific communities, as well as to the diversity of positions adopted by philosophers enquiring about their ontological status. Given that the ultimate question of the ontology of models is dismissed as meaningless, the communitarian epistemologist's job may well be to look at the different communities using models and attempt to broker agreement between them regarding an answer (any answer) in order to optimise their use.

Having sketched some aspects of communitarian epistemology as regards the topic of scientific models, I will now sketch some more general features of communitarian epistemology, before returning to the discussion of models as related to science education.

3.3.3.5 Some features of communitarian epistemology: metaphysics and ontology

In this section I will elucidate the philosophical position of communitarian epistemology, by way of sketching some alignments as well as disagreements with other philosophical views. I will begin by discussing some alignments and divergences that communitarian epistemology has with realist doctrines, thus elucidating some of the metaphysical assumptions of communitarian epistemology as I see it, and will then add some comments on the primacy of the ontology of social groups and communities over what we know.

3.3.3.5.1: Metaphysics

In this section I will say a few words about metaphysics, more specifically about three metaphysical positions, those of metaphysical realism, internal realism and promiscuous realism. I will begin with the view that is most hostile to communitarian epistemology, that of metaphysical realism. I will take metaphysical realism as introduced (and ultimately criticised, however) by Putnam to consist of the following three theses:

(i) the world consists of a fixed totality of mind-independent objects,
(ii) there is exactly one true and complete description of the way the world is
(iii) truth involves some sort of correspondence (1990, 30)

The communitarian epistemologist takes issue with all three theses of metaphysical realism. Firstly, talk of a fixed totality of objects is viewed as incoherent from the point of view of meaning finitism, a thesis that is presupposed by communitarian epistemology and about which I will have more to say in Chapter Five. The reason that a totality of objects is treated with suspicion is because any classification is considered by the communitarian epistemologist as fluid and ever-changing, hence there could not be a fixed totality stable in time. As regards the second thesis, it is rejected for similar reasons as the first, whilst the third thesis is rejected in favour of a consensus theory of truth, on which I will say more in the next section.

Hence the communitarian epistemologist is opposed to a hard-nosed realism such as Putnam's metaphysical realism. However, there is some alignment with softer forms of realism, such as Dupré's promiscuous realism. I will not proceed into a full-blown discussion of Dupré's position, focussing instead on some of the elements it shares but also on important differences with communitarian epistemology.

Dupré's metaphysical position, which he terms as "promiscuous realism" has some parallels with communitarian epistemology. Dupré claims that as regards classification of things into kinds, "many individual things are objectively members of many individual kinds" (1996, 105) and further claims that this is a combination of pluralism and realism. This view is thoroughly anti-essentialist, as it denies that there are objective properties which are the essences of kindhood and which determine the classification of things into kinds.

Dupré situates his view on what he terms as "disorder" on the metaphysical level, that is he claims that the world as such is indeterminate and the non-existence of essence is a mind-independent fact. The communitarian epistemologists align themselves with Dupré on the issue of anti-essentialism and the indeterminacy of the world, however they are suspicious of pluralism and its associations to realism on separate grounds. Briefly, pluralism sounds like an attractive position to adopt, however there is a suspicion as to whether it is a stable position or whether at the end it collapses to either relativism or hard-nosed realism. As communitarian epistemology is an epistemic position, the communitarian epistemologist would challenge the promiscuous realist on whether he can still hold on to his realist credentials given the incompatibility and perhaps downright contradiction among statements from different conceptual schemes. The pluralist answer to this, that perspectives are partial evidence of a mind-independent reality could be dismissed by the relativist as unjustified.

I will now move on to expressing some thoughts on the ontological commitments of communitarian epistemology and how these serve to condition epistemology.

3.3.3.5.2: Ontology and epistemology

Introducing a line that I will develop further in the next chapters, I will here introduce the sole ontological commitment of communitarian epistemology, a commitment which serves nicely to also ground our epistemic condition. This commitment is the commitment to communities of people armed with language. Coupled with the observation of community diversity, that is with the observation that the life experiences of other people differ to ours, this commitment proves a sufficient ground on which we can get a lot as regards our epistemology. In fact, we can build our epistemology on this basis.

In this context then, and given the assumption that we do not have direct access to reality unmediated by language, but do have direct access to other people's agreement as expressed through language, agreement and consensus play the most fundamental role. To a certain extent, as regards especially the role of language, I share Searle's (2006, 54) claim that "Language is the presupposition of the existence of other social institutions in a way that they are not the presupposition of language. " However, I disagree with Searle's position that the institutional reality presupposes a bedrock of mind-independent "brute facts" (1995, chap. 1). My claim is that the view that sees institutional and social reality as being more complex than the brute reality of the external world gets things back to front. By this I mean that rather than trying to reduce social reality to a part of the absolute external and mind-independent world, we should be recognising that the end result (the understanding) is a product of enquirers grounded in communities that function by agreement on the given topic. Simply put, my claim is that as regards what we can know, we have nothing beyond community agreement as expressed through language. In this I am aligning myself to a certain extent to Habermas' theory of truth as ideal justifiability, whereas the "truth condition of propositions is the potential assent of *all* others" and furthermore "the universal-pragmatic meaning of truth…is determined by the demand of reaching a rational consensus" ([1971, 89] cited in [Bohman and Rehg 2011]). Where I differ with Habermas in this respect is that I believe that there may be more than one endpoints of rational consensus, that in a sense rational consensus is open-ended and contingent upon the interests of the inquirers. Viewed in this light, rational consensus does not necessarily reflect a regularity in the outside world, but rather reflects the identity of interests and values of the enquirers.

Going back to the ontology of communities of people and how this conditions and to a certain extent grounds our epistemology, it is worth saying a little bit more about such communities. Such a study of communities of people should necessarily be a historical and sociological endeavour, hence I will only provide a few pointers rather than a full-blown study. Historically speaking, the primal sort of communities should be linguistic communities, thus reflecting the constitutive role of language in conditioning epistemology. Another feature of communities is that they tend to fragment and divide as they become more sophisticated and more numerous. This may be down to natural drift of interests, or in difference of the life experiences of the enquirers. By this I mean that the more numerous a community becomes, the more difficult it becomes for all its members to share the same experiences and to have the same values and interests. Because of this divergence, they begin to break the consensus on knowledge claims, thus leading into the formation of new epistemic communities. Nested communities (communities within communities) also become a possibility, depending on whether alignment is on issues judged as important and defining for the whole initial community or whether they diverge on marginal claims.

Having elucidated a bit some of the philosophical commitments and features of communitarian epistemology, I now turn back to the discussion of models within science education.

3.3.4 Models in science education

In order to look into models used in science education, I will adopt Gilbert and Boulter's (1997) classification of models as mental, expressed, consensus, scientific, historical and teaching models. My claim is that proposals that adopt a relativist account of models with an emphasis on authorised games and representations as props supplemented by publicly adhered to principles of generation can aid school-children to construct and manipulate consensus models, something which would allow them to move into the second and third levels of understanding, as described above. I will now proceed to explain this claim.

Gilbert and Boulter's classification of models comes from within the context of science education. According to this classification, a mental model is a cognitive representation of a phenomenon or object (Erduran and Duschl 2004, 117). An expressed model is a mental model that has been expressed via speech, writing or action. A consensus model, on the other hand, is an expressed model which has been subjected to scrutiny by a social group. This group may be students in a classroom or a relevant scientific community. A consensus model corresponds to Walton's (1990) authorised game of make-believe, in that its principles of generation are explicit and accepted as well as adhered to by all. A scientific model is a model that is used in current scientific practice, whilst a historical model in the context of science education is a consensus model whose use has been superseded by more accurate scientific models, and finally a teaching model is a simplified version of a scientific or historical model, used for teaching purposes.

There exists a link between the laboratory and the science classroom, namely that what is desired in both is consensus models. Such models are models that have been subject to public scrutiny and thorough testing by the relevant community. Hence, there is an irreducible social element in learning to use and reason with models. In my proposal for science education, which I will now give, it is this social element of knowledge that I wish to focus on.

3.3.5 A proposal for science education

My proposal for science education, and especially chemistry education, is to move away from content teaching regarding how nature works. Instead, students ought to be actively encouraged to construct their own models and submit them and their reasoning behind them to public, democratic scrutiny inside the classroom. At the same time students ought to be pushed to reflect upon what interests and thoughts, other than the description of the reality of the natural world, prompt them to use the analogies and models used. Such interests may include ease of understanding, or understanding relevant to a certain action.

I began this section with a challenge to the representation of the water molecule, in order to avoid the trap of taking models as literal representations or snapshots of reality. My claim is that what is ultimately desired from science education is twofold: to give future citizens an understanding of how real scientists function and how knowledge is produced in scientific settings and what motivates scientific thinking, and to furthermore encourage them to question each other in a democratic manner. Current practice, where models and analogies are often introduced in a matter-of-fact, authoritative manner, results in confusion, especially when school curricula sometimes ask students to use conflicting or even contradictory models without at the same time giving any insight into the motivations behind each model or behind its intended explanatory role. Such a behaviour sometimes leads students into confusion, such as when

Harrison and Treagust observe that

> Some of the models used in different parts of the science syllabus are even contradictory; for example, the use of conventional current (a flow of positive charge) in physics clashes with the flow of negative electrons used in electrochemistry. Maybe we should be more surprised when students are not confused by this behaviour! (2000, 1021)

In order to achieve this, it may be useful to split disciplines like secondary-school chemistry into the following two subdisciplines: the "chemistry forum", where students engage in group-work to produce models given certain prompting of phenomena under study as well as certain guiding interests that may be of social nature, and the other will be the "chemistry museum" in which students will look at historical models and the arguments and empirical evidence that led to their demise.

The first subdiscipline, what I have called the "chemistry forum", builds upon a proposal described by Woodruff and Meyer (1997). The proposal is to split the teaching time between discussion in small groups (intra-group discourse) and discussion for the whole class (inter-group discourse). Woodruff and Meyer argue that the two modes of discourse are different and that they simulate real scientists who are members of two kinds of communities, their small laboratories and the larger scientific community of their specialisation. It is argued that intra-group discourse is co-operative and distributed, whereas inter-group discourse is competitive and uses various rhetorical devices in order to persuade the audience of the validity of their views. During intra-group discussion the focus is, using Walton's pretence theory, to devise a set of rules that seems to be the most interesting or the most stimulating given a set of props, which may be common to all groups. During inter-group discussion, the focus is to discuss and argue for the merits of different sets of rules as proposed by different groups, and maybe add props depending on the consensus of the whole community. The whole situation is viewed as contingent, since the introduction of rules or props depends on the relevant community and its values, which is either the smaller group for rules to be tested or the larger group made up of the whole of the class in the case of the introduction of props.

In such an activity what is focussed upon is not the correct explanation of a phenomenon or a series of phenomena, rather adaptability to given props and principles of generation and ability to influence and negotiate one's group into privileged outcomes. Furthermore, given the predominance of the argumentation and the group work, as well as the mixture of the props, through the provision of props both from the natural and the social world, science is viewed primarily as a social activity by interested agents that is sometimes guided by the quest for truth, and at other times by other values such as coherence or peace at the social level. What is sought for is consensus itself, rather than the "truth" of any single model.

Woodruff and Meyer describe the mechanisms by which knowledge is produced in such settings as mutual knowledge, convergence and coherency. During this process of discussion all three mechanisms identified by Woodruff and Meyer come into play, as mutual knowledge comes to the forth mostly during the intra-group discussion, whereas convergence and coherency mostly operate at the inter-group dialogue stage.

The role thus of the teacher would be to co-ordinate the discussion and to enable all students to have their say in a democratic manner, and prompt the class into model-construction by giving them constraints such as principles of generation, motivating interests or props to take into account. The teacher may also experiment with different themes at different sessions, with weeks being labelled as "revelation" week (only one group has the power to impose principles of generation whilst the others have to find suitable props to test it), "truth" week in which the teacher gives unpalatable principles of generation and each group a certain social identity attached to certain values and interests, a "tradition" week in which different restrictions are given to each group in terms of the possible generation principles they may come up with, and so on.

Such a method of teaching does not focus on teaching "the right" or the "true" theory, since truth is only one among many principles adopted, whilst the whole exercise is viewed as a game, with challenges being set up by the teacher. The scientific aspects of the game can be enriched with philosophical discussions through the addition of suitable props, given that props are not restricted to phenomena from the natural world only.

By encouraging such practice, and by stressing the proliferation of explanations and of explanatory models, students may achieve the third level of understanding of models as described by Grosslight et al. (1991), that is, by building their own models with minimal constraints and by being aware of the interests and constraints driving model generation, they will come to appreciate that model construction serves primarily to analyse and develop their ideas, rather than providing the ultimate word on aspects of the world. Furthermore they will come to actually carry out the democratic (and scientific, as mentioned above) ideal of deliberation and argumentation in a democratic setting.

The second subdiscipline which school chemistry would consist of is the one I have named "the museum of chemistry". This will consist of the study of historical models of chemistry and of active re-enactment of the experiments and the argumentation that led first to the establishment of consensus models among scientists at certain times, as well as to their demise following the introduction of better arguments and further experimental evidence. Chemistry is well suited for such a historical re-enactment, since it has theories like the phlogiston theory which offer radically different ontologies and other topics such as the various models of the atom, which are suitable for a historical treatment.[8]

Of course, the above proposal shapes a different role for the instructor of science education. The teacher would have to give up their role as the authority figure that judges right and wrong in the classroom, in order to exchange it with the role of the co-ordinator of group discussions, the "elder" figure which safeguards that the discussion is guided by participatory democratic ideals and the person who chooses the appropriate phenomena to be used as prompts in order to fire up the students' imagination. Teachers would need to be at least conversant with current scientists and familiar with some of the technical literature on current as well as historical science. Scientists themselves would benefit from such a shift in science education since the product of such an education would be a critically minded student aware of the limitations of scientific modelling practice and in a way the whole of the population would be moulded into the way scientific knowledge is produced. Furthermore, the problems generated by the misunderstanding of the scientific endeavour and of the uncertainty, such as the ones highlighted in Section 3.3.2, would subside, in favour of a more engaged public as regards scientific issues.

One potential drawback of this proposal, to which I will next turn, would be that students and hence future citizens may feel less safe and confident about the physical world that they inhabit, when the comfortable link to an external reality is severed and the teaching of the predominance of interests and values in scientific activity dominates. This is akin to the worry of cynicism and turning away from science as elaborated earlier in this chapter. I now briefly turn towards this potential problem, and its democratically inspired solution.

3.3.6 Intellectual safety and science education

An objection levelled against the practice of letting students experiment with the construction of gerrymandered models and focussing only on peer persuasion and acceptance is a moral concern that such a

[8] This reflects closely the call for what Chang terms "complementary science", see Chang (2004, Ch. 6)

schooling makes sceptics out of future citizens. A worry about intellectual safety is raised, in that it is claimed that the role of school is to give future lay-people the confidence of certainty and a sense of "being in control", in order to allow them to negotiate their lives in a technological world. The argument as I perceive it would thus go as follows:

If citizens are certain about the world they live in, then they are more likely to be good citizens.

School science education gives future citizens a certain amount of certitudes about the world they live in.

Hence, school science education promotes good citizenship.

In other words, the argument is one regarding intellectual safety, which, as defined by Schrader (2004) has two elements, the epistemological and the moral. The epistemological element is the extent of the "epistemic fit" between the epistemology of the students and that of the teacher, and the moral element is the moral climate in the classroom. The argument then would be that an environment of intellectual safety is an environment accompanied by certitudes about the world based on a preferred method, that of reasoning with evidence, as is the way of science. As described by Schrader (2004, 89),

> Instructors are seen as the keepers of information and knowledge, and must do their best to impart that knowledge on the students. The role of the student is to absorb and share what they learn from professors. The nature of knowledge is that knowledge is known, certain and knowable—at least by well-qualified authorities or by authorities who are working on finding the answers to specific problems. Factual information is perceived as infallible, evidentiary, and evaluated by external authority for its verification. If knowledge is unknown, there is a sense that there is a correct method or approach to finding it, and students learn that if they cannot learn the knowledge itself, they will learn from the professor the method to use to attain the answer, and will be evaluated and judged in how correctly they master this.[9]

Schrader makes a slightly modified claim to the one made in the argument I have laid down in the beginning of the section, as the claim is about a method that can be used to yield the answers sought for, if these answers are not actually available. If this refers to the scientific method, then Schrader's argument falls prey to the rejection of a single scientific method as laid down in Chapter One. However, I will treat Schrader's

---

[9] It is worth noting that Schrader is very close to mirroring Kuhn's views on education in normal science, as described in his essay "The essential tension" (Kuhn 1977).

thesis as broadly lending support to the argument as laid down earlier.

The environment described by Schrader is opposed to what Perry (1970) calls relativistic environment, where the contextual and historically contingent character of knowledge is brought to the forth, and where authority shifts from the teacher to the students, and the teacher ceases to be the bearer of what is "right".

My answer to the argument from epistemological safety as laid down above proceeds on two fronts; first I will argue that both premises are false, and secondly I will argue that even if both premises were granted, then the conclusion would still not follow as school science education as the provision of certitudes is simply irrelevant to good citizenship.

To begin with the examination of each premise, the first premise can be shown to be false if one considers religion as a source of certainties about the world. The move, is, I think, a valid one, as, often one hears religious leaders claiming that religion provides believers with absolute certainties about the world. However, one would be hard-pressed to claim that religious people are better citizens than non-religious people. Indeed, in the beginning of the 21st century, religious fundamentalism is seen as a threat to the foundations of the modern democratic state, which, it is claimed, is based on secularism, the sharp division between Church and State, and tolerance of different religious affiliations. Hence, it seems that at least one source of generating certainties about the world, that of religion, is rejected as promoting good citizenship. I thus conclude that the first premise is false, in the sense that it can be the case that citizens may be certain about aspects of the world they live in, but nonetheless not be good citizens. The above argument does not depend on whether the certainties provided by religion turn out to be true or false, as one may claim that moral facts provided by religion are true, whilst one may be agnostic about facts about the natural world, such as the existence of one or multiple deities.

As regards the second premise, it is not the case that science education provides certitudes about the world, for the key reason that science itself does not provide certitudes. Neither science-in-the-making nor long-settled science claim to provide the final word on any matter, but rather provide quasi-certainties (at best) relative to the amount of evidence currently existing. The first because evidence is often lacking or elementary, the second because it is often accepted by scientists that what it provides are crude simplifications of what reality is, approximate truths rather than truths simpliciter.[10] Indeed, one could claim, in the sometimes

---

[10] This is an internal criticism, from the point of view of the scientific realist. There

bitter division between science and religion, that this is a key difference between the two, in that religion claims to be the source of absolute certainties whilst science is all about critically re-examining and abolishing dogma. Hence, if science education even purports to provide certainties about the world, then it seems to be going against its essence which is to teach science. I thus conclude that both premises of the above argument are false.

However, even if one were to accept both premises, the conclusion would still not follow, since the certitudes provided by school education are often simply irrelevant to the way citizens negotiate their daily lives and hence the actual content of the approximate truths matters little in our later lives, both as human beings and as citizens in democratic states. Very few lay-people know or need to know how telephones or computers or televisions actually work, beyond that they all use electricity and that electricity is a form of energy, same as the fuel that fuels our cars, and we all know that if we hit a brick wall then that will hurt us without knowing exactly of Newton's laws. Furthermore, according to the sophistication of our religious views, we may or may not remark the similarities between apes and humans and thus approve of the idea that apes and humans are relatives in the grand tree of life. All these bits of approximate knowledge can be derived from everyday negotiation and practice of daily life. Furthermore, issues such as the fine structure of matter or the valence structure of acid atoms also often matter very little in our daily life.

As a point of qualification however, it should be remarked here that this argument concerns the way science is actually taught and the focus on transmitting facts to students, and that I am not making the claim that all science education is useless. What is sought is a science education that focusses more on the production of scientific knowledge and its democratic credentials, rather than the content of the scientific facts disseminated to future citizens.

I thus conclude that the conclusion that science education promotes citizenship simply does not follow from linking science education with certitudes about the external world. To reiterate, being provided with certitudes seems simply irrelevant to being a good citizen, hence there is no need for intellectual safety nor the environment described by Schrader for good citizenship to exist.

---

is however also an external criticism consistent with the communitarian epistemologist viewpoint that I am advocating here, that purported scientific certitudes are relative to the relevant communities, and that they are simply disseminated into more people, who then become the new subjects of the epistemic community regarding a given "fact of the matter".

A final retort may be that there simply is a link between knowledge of truths and a moral sense of safety, what Schrader labels the moral element of intellectual safety. An answer to this would be to claim, as in the discussion of radical education, that the moral element of intellectual safety also lies in the method of co-creation and co-ownership of knowledge at the level of the community. In a sense, what is relevant is not the truth or the end product of the process of learning, rather than the method through which one reaches this end point. And this method, in order to be beneficial to society as a whole, had better be a democratic method that privileges the good of the community over individualist growth and competition. Furthermore, such method had better inform future citizens about the indeterminacy and the complexity of the external world in which they are called upon to build their lives.

In section 3.3.3 I discussed some implications of communitarian epistemology regarding realism and anti-realism, both as regards models as well as more generally. In concluding the present section, I will elucidate a bit more on a theme mentioned in the above discussion, that of indeterminacy within conceptual schemes as well as between them. In the context of communitarian epistemology, a discussion on intellectual safety is central, for the main reason that if a solitary knower can possess intellectual safety through unmediated interaction with reality, then there is no need for perspectives such as the communitarian one. However, the indeterminacy that arises in philosophical questions or foundational issues of the sciences, or in something as simple as the Sorites paradox directly threatens intellectual safety for the solitary knower. In such a situation, safety arises out of strength in numbers: a discussion ensues with its endpoint being either consensus or an opinion shared by the majority. Hence intellectual safety is restored through democratic deliberation.

3.4 Concluding remarks

In this chapter I have sought to link the topics discussed in the previous two chapters, namely the role of scientists and citizenship in a participatory democratic society, as discussed in Chapter One, together with the perceptions of science by lay-people, as discussed in Chapter Two. My avowed aim was to stress how citizenship and science education can be linked and reinforce each other. In order to achieve this, I identified the common thread of argumentation that is and ought to be treated as constitutive of both. I first identified the small tradition of radical education that goes back to the writings of John Dewey as an example of participatory education in action. I then proceeded to discuss whether science and citizenship are truly compatible before highlighting the notion of argumentation as constitutive of both.

I then embarked on a case-study of the topic of the use of models in education and some related problems with models within the domain of the philosophy of science. Whilst the initial impetus of the case-study was to highlight the problem of the misunderstanding of the process and products of scientific modelling within the classroom, I also used it in order to clarify my position on the realism versus anti-realism debate on the topic of the ontology of models. This was deemed necessary in order to elucidate certain features and commitments (or lack of) of communitarian epistemology, which is the philosophical position that I will defend more rigorously in the following chapters. Moving away from the discussion of models I also moved to elucidate some of the more general features of communitarian epistemology, regarding the relationship between ontology and epistemology, as well as its metaphysical commitments.

The foray into the philosophy of scientific modelling also served to justify what follows, namely my proposal for a shift away from content teaching on the topic of models within school science education and into the process of producing consensual models through public argumentation within the classroom. This is because the focus of communitarian epistemology is community agreement in order for knowledge to be co-produced and co-owned by the community of enquirers that is the classroom, rather than knowledge being transmitted by the teacher to the students. In this connection I also related the sole ontological commitment of communitarian epistemology, that to communities of enquirers armed with language, and elaborated on how this may serve to ground our epistemology.

I finally addressed an objection to the relativist epistemology that I propose, in the face of an argument for intellectual and moral safety in the classroom. I concluded that intellectual safety does not have to be linked to the provision of certitudes and that in any case such safety may be provided through teaching based on consensus or democratic deliberation.

In this chapter I have begun answering the questions posed in the first two chapters, on the problem of scientific elites. I have sketched a route into a democratic polity, by focussing on education and in particular how science education can be used to promote democratic citizenship through the notion of argumentation. I have also begun gesturing towards a philosophical account that would incorporate and promote democratic citizenship in the face of communitarian epistemology. Such an epistemology combines well with democracy as it eliminates the need for truth as grounding intellectual safety, replacing it with community consensus and strength in numbers, which is a key notion of democracy.

In the following two chapters I will focus more on defending communitarian epistemology philosophically. I will first begin by defending communitarian epistemology as an epistemology that gives priority to the community rather than the individual and at the same time describes accurately certain aspects of scientific practice as well as decision-making in or based upon science, before moving on to the fifth chapter to defend meaning finitism, a theory of meaning which complements communitarian epistemology.

Appendix

A Proposal for Science Education in Primary School: Demonstrating and Myth-Making

In this section I will proceed into a proposal for the teaching of science to primary school children, continuing on the theme of putting some flesh into the notion of promoting science as primarily a social activity.

A main problem in teaching science to primary school children is the acquisition by them of concepts for which it is hard to provide demonstration. Concepts like atom, mass, energy and time are hard to understand, not least because they pose philosophical as well as physical problems for grown-ups, let alone for primary-school children. Another problem with these concepts is that the simplified ways in which children are taught about them are often in dissonance with the consensus among scientists as regards their use. Of course, one may argue that children lack the conceptual apparatus to understand the complicated definitions of concepts made in the language of theoretical physics.

My claim is, however that the woolliness of concepts and lack of clear and simple definitions used in disciplines such as fundamental physics is a weakness of the discipline and the way the activity of science is conducted rather than being a problem in the minds of lay-people. In other words, contra Wolpert (1997) I claim that fundamental science should not be difficult and should not appear difficult to lay-people. In a sense, my claim is that what scientists claim they are striving for, such as simplicity and beauty (see for example (Kuhn 1962) for the identification of these virtues in science) should be evident to everybody at first sight rather than only to the initiated, and if that is not the case, then the fault lies with the scientific community.

Of course, the claim is not that a child should be able to understand every technical detail of how the Large Hadron Collider at CERN works, but rather that lay-people and even children have an accurate and congruent with current scientific wisdom grasp of what concepts such as "energy" and "light" mean.

My proposal then for primary school education is that the science curriculum for primary schools should be divided among two main lines of knowledge-transmission, and that these lines should concern the demonstrative and the speculative elements of science and technology. The demonstrative and interactive element of science and technology is the creative part of science, which will allow the children to engage and experiment with simple technological innovations. Examples may be the construction of water-filters, pH colour scales, etc.

Emphasis should be given on the simplicity, and where possible use of household ingredients, in the experiments to be performed. The aim of such a "do-it-yourself" approach will be to demonstrate the promiscuity and omnipresence of science. Hence, the message will be that science is and can be done by anyone. The teaching may focus on examples where simple ideas in the sciences have proven to be of much strength in explaining phenomena, and they may also combine elements from many different disciplines, such as applications of chemistry and physics in biology or applications of mathematics and geometry in physics. Room should be made for local-traditional solutions to every-day problems such as the conservation of meat or cold-resistant architecture and insulation. Through such teaching practices, however there should be a conscious effort of moulding school-children into both "low-level" science, that is science that is done in the house, and "high-level", that is science and technology that is done at specialised laboratories. In a way, the "black box" (Latour 1987) of technological innovations should also be attempted to be partially opened to primary school children, rather than making a sharp separation between the "difficult" high level science and technology and the "easy" low-level science.

The second part of the curriculum is the speculative element, which, as was hinted above, would have as its primary focus the introduction of school-children to the main ideas motivating present research into theoretical and hard-to-demonstrate elements of science. The main motivation of this is that rather than follow an approximative route towards some deep concepts about which there is confusion, to directly inform children of surprising elements of ideas currently accepted. To demonstrate the difference between the approximative route currently preferred and the more direct route that I am proposing I will use two examples, of which the former is the transmission of a strictly speaking false belief.[11]

Two examples of belief-transmission through myth-making

The first claim is the claim that the Earth is a sphere circling around the sun. This claim would appear counterintuitive to children of primary school, who would naturally assume the flatness and immobility of the Earth. The further problem that I would identify with this claim is that, being an approximate truth and a truth of long-settled science, it is at odds

[11] Actually, an "approximately" true belief.

with the actual opinion of scientists, who accept that the Earth is a flattened sphere having an elliptical orbit with the centre of the ellipse not coinciding with the centre of the sun. Hence I conclude that the promotion of the roundness and of the Earth and of the circular nature of its orbit should not be given as much focus as it is given, but rather that a different approach should be followed in transmitting to children something which is closer to the opinion of present-day scientists on the matter.

A modified speculative claim would be to focus on the non-flatness of the Earth and straight-away teaching of children about the elliptical orbits and the real shape of the Earth. This could be done easily by showing real photographs of the Earth taken from space and of the simulation of the orbit of the Earth around the Sun using real data.[12] A further, speculative move to demonstrate the mobility of the Earth would be to attempt to show through specially designed children's stories how it may appear to one that they are immobile but to be actually moving, in order to show to the children through a speculative medium but close to their experience, how we currently think that the Earth is moving. My claim is that since what is being proposed is counter-intuitive anyway, the current cutting-edge opinion would be a good place to start for the transmission of knowledge.

The second, this time true claim that could be projected in a speculative manner is an effect from quantum mechanics, the quantum tunnelling effect. This is an effect from more cutting-edge science, and one that can stir the imagination of children, since it appears that electrons are jumping through the equivalent of brick walls. Hence, a childrens' story can be constructed around the message that it is possible for some special "magic" pieces of each and every one of us to pass through walls, and then ask children to entertain the possibility that such stuff which appears to be a part of a fairy story is actually an element of reality. The last step in this would be to introduce to children images taken with the use of scanning-tunnelling microscopes and have perhaps specialists working on such images make a presentation tasked to impart some of the "magic" of doing experimental physics to primary school-children. Such a test would benefit both parties, since it would force the scientist to think hard about making their presentation understandable by small children whilst not straying away from current scientific orthodoxy. Furthermore, such a bridging of the gap between experimental physicist and primary school children would have the effect of giving intellectual safety to school-children, since they

[12] Though care should be taken not to fall into the trap of believing that the simulation is the final word on the matter, see the section above on models.

will no longer be intimidated by "hard" physics, but rather be at ease when confronted with ideas from that realm.

It is worth stressing that all manners of scientific truth could be transmitted to children in this way. An example of a claim introduced in a mythological manner would be to ask the children to imagine that they are riding a light ray, and tell them that nobody could catch them if they did so.[13] However, this would be countered by them being very massive, in the literal sense of the word.

It is worth noting, following Feyerabend (1975), that scientific claims should not be privileged in the teaching of very young children, except by pointing out the fact that they are believed by many people educated in the West. It would be beneficial if other claims such as religious claims about the origins of the world are at least presented in the same manner to children, in order to stimulate their imagination. Of course, one problem that this leads to is the problem of the goals that science education in primary school should have, and especially its continuity with secondary-school education.

Conclusion

The aim of this brief proposal has been to give an indication of how science can be taught as a blending of myth and reality to primary school children, in order to help demonstrate that science is primarily a social activity that stimulates the collective imagination of the participants and challenges it with interesting phenomena, rather than a dry presentation and discovery of facts about a mind-independent world. Furthermore, it is designed to show how in that regard science as conducted in universities and other institutions in the 21st century is not that different to other cultural activities practised outside the boundaries of academia or other "scientific" setting. A further target has been to challenge the approximative route towards learning scientific "facts" and the assumption that school-children have difficulties forming concepts and that thus a simplified approach should be followed in their education. I have presented, through the example of the motion of the Earth, how the teaching of an approximate (but ultimately false) truth taught currently in schools can be replaced by the teaching of beliefs more congruent with those of current scientific wisdom, and I have argued for a pluralistic

---

[13] This is actually one of the pillars of special relativity theory, and a thought experiment that played a large part in Einstein's formulation of the speed of light postulate.

education which will consist in the introduction of the prevalent ideas in science to school-children from an early age, as well as the presentation of other technological innovations not falling under the scope of "science".

CHAPTER FOUR

COMMUNITARIAN EPISTEMOLOGY AND SCIENCE

4.1 Introduction

The previous chapter ended with the identification of communitarian epistemology as an epistemology that promotes the community as the primary unit of analysis in the search for the epistemic subject, and that it furthermore promotes knowledge as the result of argumentation and consensus. In this chapter, I will defend the two main theses of communitarian epistemology, as they were laid down by Kusch (2002). Given that previously I highlighted the need to arm lay-people with an epistemology which accurately describes the production of knowledge in scientific settings, it will be of no surprise that the line of defence of the two main theses of communitarian epistemology presented here will be to argue that they accurately describe aspects of the production and dissemination of scientific knowledge.

Communitarian epistemology consists of the following two main theses:

- The primary epistemic subject is the community
- Knowledge and its cognates (such as "knower") designate a social status (see Kusch (2002, 1))

In the first part of the chapter (section 4.2) I will defend the first thesis; I will do that by arguing that for an individual to know a claim P means being part of a community that knows claim P, at least as regards scientific claims; in the second part (section 4.3) I will defend the second thesis, by laying out possible sets of entitlements and commitments that members of different epistemic communities may have towards a knowledge claim. It is worth remarking right at the outset that whilst the second thesis covers all cases of knowledge claims, the first thesis may be violated in certain specific situations, such as Robinson Crusoe cases or the case of a scientist coming up with a theorem but not having the chance of telling anybody before they die (see (Kusch, 2002, 1)). However, it is arguable whether these cases put the thesis itself into danger or even whether they are qualitatively different to other cases. This is because of the strong thesis that communitarian epistemology takes of normativity[1], which rules out solitary rule-followers. In light of this thesis, Robinson Crusoe cases are to be seen as limiting cases of the community thesis rather than as violations of it. They are simply arcane cases which are parasitic to more orthodox cases of scientific knowledge acquisition.

In the second part of the chapter I will defend the second thesis of communitarian epistemology, which states that knowledge and its cognates designate a social status variously attributed to communities ("knows"), individuals ("knower"), and propositions themselves ("is known"). My strategy in the second part will be to first present a sociological phenomenon taking place within science along with an explanation of it from within the literature, and then criticise this explanation and in turn suggest a communitarian epistemological explanation which, according to me, explains the phenomenon better. I will conclude by laying out different entitlements and commitments that different communities may have as regards a given scientific knowledge claim, in an effort to put some flesh into my proposal.

4.2 Community versus the individual as the primary epistemic subject

The first issue that I will tackle will be the issue of the primary epistemic subject. The thesis which I will be defending is the following:

An individual X obtains knowledge of scientific claim P in virtue of being a member of a community A that regards claim P as knowledge.

It should be remarked that this claim includes the originator of claim P, that is the first scientist that will claim that P. In other words, what the thesis states is that a claim P only becomes scientific knowledge once it has been through a process of validation by a scientific community. Hence a four-step procedure may be described, consisting of the following events in chronological order:

- A) Individual X asserts scientific claim P
- B) Individual X proposes scientific claim P to another person or persons
- C) Community A, consisting of individual X and at least one other

[1] Which I will discuss at section 5.6

person, agrees (subject to the community's rules[2] of justification or proof) to treat claim P as knowledge at which instant it follows that
- D) Individual X, as well as the other members of community A, know claim P

The above-sketched picture is meant to be contrasted with the following two-step picture:

- A') Individual X obtains knowledge of scientific claim P through perception or inference
- B') Individual members of community X obtain knowledge of scientific claim P each for themselves through their own cognitive capacities supplemented by X's testimony

The strategy that I will follow in arguing for the conclusion that the primary epistemic subject is the community will be the following. In a first instance I will argue that the thesis applies to scientific knowledge production, arguing that in what is widely recognised as scientific activity (the activity which claims as its originators the methodological writings of Bacon and Newton) the scientist is never alone, even if they are the single author of a scientific work. I will draw on certain insights from Latour's (1987) study in the making of scientific knowledge to support the thesis that the individual scientist is necessarily surrounded by allies. However, as opposed to Latour, I will only focus on human allies.

This attempt will consist of two parts, the first being the exploration of what I term the intra-laboratory aspect of scientific activity, and the second being the public forum aspect. I will conclude that the latter aspect is the stronger aspect in which a claim that the production of scientific knowledge in principle contains an irreducible social element, that is that the appropriate unit of epistemological analysis of the production of scientific knowledge is the scientific community rather than the individual scientist.

In a second instance I will conduct an analysis of some of the individualistic shortcomings of traditional epistemology, claiming that the focus on linking mental states to the outside world as integral to epistemology has resulted in ignoring social aspects of epistemology.

Finally, I will elaborate and discuss the thesis that community agreement is constitutive of knowledge, presenting and arguing for the

---

[2] These rules are of course, different from group to group and even within the same group at different times).

communitarian account of scientific knowledge (Kusch, 2002). I will now begin my account by arguing that the community is constitutive of the production of scientific knowledge.

4.2.1 Presences in the laboratory and in research papers

As Ziman remarks, (2000, 4), arguing that science is a social institution,

> It (science) involves large numbers of specific people regularly performing specific actions which are consciously coordinated into larger schemes. Although research scientists often have a great deal of freedom in what they do and how they do it, their individual thoughts and actions only have scientific meaning in these larger schemes.

This quote seems to accurately describe the mode of work of scientists working in large "collaborations" such as those encountered at CERN and described by Galison (1997) or by Knorr-Cetina (1995).

In the above, two senses in which science is a social activity may be discerned. The distinction between these two senses concerns two aspects of scientific practice, what I will term the intra-laboratory aspect and the public forum aspect. In what follows I will elaborate on these two aspects. I will conclude that whilst the first aspect is not sufficient to prioritise the group over the individual as regards the production of scientific knowledge, the second one can fulfil this role.

4.2.1.1 The intra-laboratory aspect of science as a social activity

The intra-laboratory aspect of science as a social activity focusses on the interactions that take place among individuals at the time of their induction into the sciences and during their day-to-day laboratory work. These interactions may range from interactions with specialised technicians and helpers, as in the case of Boyle and his French assistant Papin, or Hooke (Shapin 1994, ch. 8), to the specialised division of labour taking place in bigger experiments. It is worth remarking that specialization in science and division of labour took place early on in science, at the same time as it was becoming part of academia in nineteenth century Germany (Ziman 1976, 229). Such an arrangement, which was essentially a master-apprentice relationship, had as a result the formation of a "school of research", where the students were working on problems selected by their professor, adopting their methods and opinions. It is worth remembering that even scientists purported to work "on their own" or who published on their own, usually worked with assistants[3].

My claim is that scientific activity has always been a group activity, rather than this being an accident of 20th century practice. However, this claim does not preclude the possibility that certain lines of inquiry could at least in principle be open to the individual scientist. Furthermore, as regards the production of scientific knowledge, it can still be claimed that the individual that heads the scientific laboratory or is the first author in a scientific paper still has priority over the rest of the authors or the collaborators. The head researcher is necessary to the production of scientific knowledge as described above in a way that the assistants are not. I will, however, now move onto the second aspect of scientific work, what I have termed the "public forum" aspect, which I claim is constitutive of scientific practice, and that necessarily links the production of scientific knowledge with scientific communities.

4.2.1.2 The public forum aspect of science

The second aspect of science which I believe necessarily involves an appropriate community into the production of scientific knowledge is that it is primarily a conversation among peers, a dialectic process in which scientists attempt to convince their peers that their representation of nature is the most accurate one. In a real sense of the word, even when Isaac Newton was locked away in his Cambridge study, he was engaged in conversation with Gottfried Leibniz on infinitessimal calculus or Robert Hooke on the nature of light. Ziman, in arguing for the constitutive role of conversation in science, states that

> Science, by its very nature, is a body of public knowledge, to which each research worker makes his personal contribution, and which is corrected and clarified by mutual criticism. It is a corporate activity in which each of us builds upon the work of our predecessors, in competitive collaboration with our contemporaries. (Ziman 1976, 90)

Hence science, even in its most theoretical guise, is conceived of as essentially a dialogue among scholars, both of other times and contemporaries. The words of Newton's famous remark "if I have seen

[3] It is telling that even though Galison portrays the emulsion expert Pierre Demers and the physicist Marietta Blau as 'lone guns' representing an older way of working within physics that was rapidly going out of fashion in the interwar years, he does produce evidence that they both worked with small teams of assistants and doctorate students (Galison 1997, ch.3)

further it is only by standing on the shoulders of giants" seem to reflect his debt to previous great minds with whom he is in a scholarly conversation.

The more proximate aspect of this dialogue is often reflected in the letters and published written articles that scientists use to communicate their findings, as well as the conference presentations and scientific colloquia that bring together specialists from all over the globe. The institution of writing in journals[4] also helps formalise the argumentative essence of science. Scientists present their work in a public forum and thus make it available for criticism and refinement. They engage directly with whoever wants to research on the same subject, by challenging each other to come up with counterexamples and counterarguments that would prove them wrong. It is perhaps no accident that the activity that Boyle and Newton were engaging in was termed natural experimental philosophy (Shapin, 1994, ch.7), that it was considered as related to the argumentative mode of inquiry of the ancient Greeks, albeit with the empiricist restriction that evidence be drawn immediately through the senses and by the questioning of nature through experimentation.

4.2.1.3 Science as an argumentative activity

In the same vein of viewing science as primarily an argumentative or rhetorical activity, Latour (1987, ch. 1) rather provocatively argues that in science in action (or science-in-the-making, see Gregory and Miller (2000 chs. 6 and 10)) what essentially happens is a mustering of opinions and allies, as expressed through the citing of references in a publication. According to Latour, when one is arguing for a scientific claim in a publication, rather than being alone they are immediately surrounded by allies and potential allies, which they can mobilise to support and strengthen the points contained in the publication. Giving an example, Latour claims that to dispute a renowned scientist's claim, one would have to oppose him and his co-workers at the laboratory, the university board that gave them the funds to carry on their research or any other awarding bodies that conferred their trust upon the scientist, the journal referees and editors who selected the paper for publication, even before moving on to read the paper and look into the scientists whose claims are mobilised to support sub-points in the argument (Latour 1987, 31–38). Hence it can be concluded that a whole community of people is enrolled in the expression of a scientific claim, even if the credit goes to an individual scientist. In a way, the individual is simply the spokesperson for a large community of

[4] Which is contemporaneous with the institutionalisation of science.

people who bestow their trust upon them, and is arguing against the spokespersons of other communities of scientists who may hold opposing opinions. Furthermore, the whole process takes place in front of an attending jury audience who after examining both opinions decide, according to commonly agreed standards, on the endorsement of one or the other opinion. It is to this "audience" that I now move my attention, presenting a more formal account of knowledge production and dissemination which accords epistemic priority to the community.

4.2.2 Communitarian epistemology and collective knowledge

As noted above, in this section I will focus more on the role of the community as validating knowledge claims that individuals may present to it. Expressing the social and communicable aspects of knowledge that I have gestured towards above, Welbourne (1986, 5) advances the claim that

> Our idea of knowledge, to put it roughly, is the idea of communicable information, information as to the facts, information which is objective in the sense that it is not dependent on any particular point of view, but is available to any one at all, with the capacity to understand the utterances in which it is embodied.

Welbourne backs this claim up by arguing that one needs to contextualise the use of the concepts of belief and knowledge by acknowledging that they have evolved hand-in-hand with a certain model of communication of information (1986, 4)[5].

Expanding Welbourne's project, my claim is that there are two different projects that have been collapsed into one in most of recent epistemology[6]. The first project is the analysis of how the term "knowledge" and its cognates are used in everyday language, and what this tells us about the nature of the concept of knowledge. The second project which has taken place under the same umbrella is that of mapping mental states (such as "belief that P") into states in the world (the fact that "P" or simply "P"). Now, traditional individualistic epistemology, influenced by the Cartesian project[7], has tried to find answers that advance both projects at the same

---

[5] In this connection it is worthwhile mentioning the work of Edward Craig (1990) who claims that the term "knowledge" is applied to the testimony of individuals who are considered as reliable sources of information.

[6] I thank my supervisor, Prof. Alexander Bird, for helping me make explicit this line of thought.

[7] Which considers a solitary thinker faced with and trying to make sense of the outside world.

time by tailoring linguistic analyses into fitting analyses of mental states corresponding to the outside world. Positions such as reliabilism (see for example Armstrong [1973]) and the causal theory of knowledge (Goldman 1967) are prime examples of this conflation of linguistic analysis and mapping mental states onto the world.

It is, however, useful to bring the two projects apart, especially if one is to reject the Cartesian model of the solitary thinker faced with the world and replace it with a model of communities of like-minded individuals trying to communicate information (about the external world, as well as about themselves) to each other. If such a move is granted then the second project of mapping the mind/brain onto the world becomes of secondary or none at all epistemic importance, and the linguistic project of analysing how the term "knowledge" is used becomes the proper and perhaps sole domain of epistemology. The question then of "grasping that P" (as described in the two-step procedure sketched in section 4.2) still remains a useful question to ask, and one that I will try to reconcile with the social picture in the account I will give below of plural subjects of belief. However, the project is no longer an epistemic one but rather one that may fall under a different label.

In the light of the above separation of the two projects, I can now examine a bit more what I consider to be the root of the problem, which I have identified as the individualist tendencies within epistemology.

4.2.2.1 The root of the problem: knowledge as individual mental state

Traditional individualistic accounts of knowledge, springing from the Cartesian model of the lone thinker, tend to ignore Welbourne's point about the significance of communication in knowledge acquisition. This is further enhanced by the reductionist position as regards testimony as a source of knowledge. This position takes testimony not to be among the standard basic sources of knowledge, which are taken to be memory, perception, reason and introspection. Modern proponents of such a position are Van Cleve (2006) and Audi (2006). Supporters of this position focus their criticism on arguing that the justification of testimonial claims has a non-testimonial basis (Goldman 1999, 126).

Two further claims add to the downgrading of testimony and of communication in general as regards knowledge acquisition. The first is the idea that knowledge is somehow "denatured"in the process of communication (Welbourne, 1986, 49). Welbourne explains that the received view claims that there is some loss of understanding of some crucial elements required for a claim to count as knowledge during the transmission from speaker to hearer. Welbourne gives the example of John Locke (1975) as an early proponent of such a view[8].

The second is the insistence on knowledge being primarily resident inside the brain/mind, that is of knowledge being primarily a mental state. This assumption seems to go unchallenged and taken for granted by most epistemologists. A particularly pertinent example is social epistemologist Alvin Goldman, who, even though he sketches his project as seeking "social paths or routes to knowledge" (1999, 4), he intends this to mean that "considering believers taken one at a time, it (his project) looks at the many routes to belief that feature interactions with other agents, as contrasted with private or asocial routes to belief acquisition"(1999, 4). He furthermore elucidates this by maintaining that "Even in this second perspective, (the social one), however, the knowing agents are still individuals". (1999, 4)

The commonly-held view is that possessing knowledge is a mental state, either in virtue of being a species of belief or in virtue of being a separate but nonetheless mental state[9], and that furthermore mental states are primarily (if not only) reducible to individual brain states. Such a thesis may be found to underlie the dialectic of epistemological writings of writers such as Dretske (1991) and is prevalent in Fodor's proposals and defence of the Language of Thought Hypothesis, in which he comes out in favour of a strong Cartesianism (Fodor 2008, 12).

Margaret Gilbert lays down two theses, which taken together explain the prevailing tendency to locate knowledge within individual's brains and preclude the possibility of group belief. These two theses she labels as psychologism about belief and anti-psychologism about social groups (Gilbert 1992, 238). The first thesis is taken to reflect the idea that only a being with a mind can have an attitude towards a proposition, such as a belief. The second thesis is meant to block appeals to such entities as group minds[10] hence it simply conveys the idea that social groups do not have minds.

A general criticism for such individualistic accounts is that they overstate the role of the individual mind by considering a lone thinker in

---

[8] Cf. Locke (1975) "In the Sciences, every one has so much, as he really knows and comprehends: What he believes only, and takes upon trust, are but shreads; which however well in the whole piece, make no considerable addition to his stock, who gathers them."

[9] Williamson claims that knowing that P rather than believing that P is a basic mental state. (Williamson 1995; Williamson 2000)

[10] It should be noted, nonetheless, that functionalists about minds, such as Williamson, would not in principle preclude the possibility of group minds.

the universe, that is they assume that individual mental states are necessary and sufficient to explain individual action, where what at best can be established, at least as regards everyday tasks, is only that individual mental states are necessary in causal explanations of action.

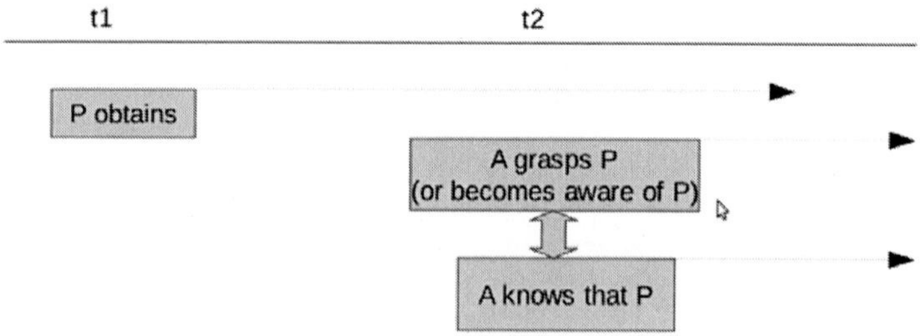

Illustration 3: "Instanta*neous" knowledge acquisition*

My argument as regards this claim is that a great number of actions can only be sufficiently explained by supplementing the analysis in terms of mental states with considerations of rule- and norm-following or other activities which have an ineliminable social element.[11] Hence I conclude that in such situations mental states alone[12] are not sufficient to explain behaviours observed

From a purely epistemological point of view, the appeal of the idea that the possession of knowledge is a purely individual mental state may be derived from the widely-held belief that it is perception and inference that are the basic sources that generate knowledge, with memory and testimony regarded as merely transmitting knowledge (however see the discussion in [Lackey 2005] for a defence of the claim that memory is a generative source of knowledge). By regarding sense perception and inference as the only basic means of generating knowledge, knowledge acquisition is

[11] A strong thesis would ground rule-following on the community, see Kusch (2002, ch.14), however a weaker thesis acknowledging that a lot of our behaviour consists of mimesis of past reactions of our peers in similar situations would be sufficient.

[12] Even states that Williamson labels as "factive", which are states (or attitudes) that one has "if and only if, necessarily, one has it (them) only to truths" (Williamson, 2000, 34) would not be sufficient to explain action taking place within a social environment. Considerations of Robinson Crusoe-like examples would be unsuitable as counter-examples as it is an open question whether congenital social recluses would reason and act in the same way as social beings would.

treated as more or less instantaneous, in the sense that an epistemic subject A comes to acquire knowledge that P as soon as A becomes aware of or grasps P, as described in Figure 3.

My claim is that this schema is an oversimplification of knowledge acquisition and does not hold for most cases where knowledge is acquired. It may be the case that an internal process of "cognition" or of obtaining a mental state of belief or a factive state ("grasping that P") does take place, and that this event coincides or precedes gaining the status of "knowing" claim P by an individual, however, given especially the thesis that the primary epistemic subject is the community, such a mental state does not play any epistemic role. In this I follow the lead of Kusch (2000, 15–38) in favouring sociophilosophy against psychologism, which is the position that the answers to major epistemological questions have to come from sociological enquiry, and that things like "reasons" and "arguments" are also social entities, just like knowledge claims and propositions. However, if one is to insist on knowledge being linked to unseen and probably mysterious mental states being linked to knowledge, then the communitarian epistemologist may concede the existence of these states and that they may play a role in causing the community to adopt a certain claim P as knowledge, perhaps as the components lying at the base of communal performatives (such as "We declare claim P as a knowledge-claim") which are themselves indispensable to communitarian knowledge-formation. Nonetheless, the linkage of facts about such mental states with epistemology is judged as a category mistake[13].

However, before proposing an alternative schema that regards knowledge acquisition as a process spread out in more time-slices than the above schema, I will briefly digress into an account of belief such that provides a concession to traditional epistemology and the conflation of the two projects mentioned above. The picture of plural epistemic subjects would also maintain the view of knowledge as justified true belief.[14]

I will hence give a non-mental interpretation of belief that would allow one to keep the traditional formulation of knowledge as justified true belief but nonetheless cast it as inherently social. This attempt is based on

---

[13] It should be noted that the argument I will present in favour of the sociological account of acquiring knowledge as gaining a celebratory social status proceeds from the observation of scientific practice and hence restricts itself to scientific knowledge, even though in principle it could be extended to other sorts of knowledge.

[14] See Rolin (2008) for a defence of scientific knowledge as collective knowledge, with knowledge meaning justified true belief or acceptance held by groups as plural subjects (in Gilbert's sense).

Gilbert's (2000) definition of belief based on joint commitment. I will then elaborate on Kusch's (2002) more detailed account of the mechanism of the generation of knowledge through testimony.

4.2.2.2 Belief as a property of plural subjects

Gilbert ([1992], [2000]) proposes an alternative account of belief that sidesteps the problems that acceptance of these two theses would cause, by proposing and defending an account of collective belief as jointly accepted view, whereby the term joint acceptance does not entail that any subset (even composed of one person) of the group consists solely by individuals individually holding the belief. In other words, Gilbert stresses that her account is non-summative, that is the joint acceptance of the view does not reduce to the sum of the beliefs of the group members (1992, 306). Gilbert gives the following formulation of her account of collective belief: “There is a collective belief that p if some persons are jointly committed to believe as a body that p. These people can then accurately say of themselves that “We (collectively) believe that p.” (Gilbert, 2000, 38)

In explaining the notion of joint commitment Gilbert stresses the fact that it does not reduce to personal commitments, and that furthermore it cannot be rescinded by any of the parties acting unilaterally (2000, 40).

There are two main advantages to be gained from adopting Gilbert's notion of collective belief and the derivative notion of collective knowledge, defined simply as justified true belief or acceptance held by groups of people considered as plural subjects (Rolin 2008, 115). The first is that it provides a definition of knowledge in line with more standard definitions in terms of justified true belief, whilst at the same time taking into account the arguments of the above sections which purport to show that scientific knowledge is the property of plural subjects. The second advantage is that in using the language of joint commitments, Gilbert is in line with the communitarian account of knowledge which I am advocating. According to this view, as elaborated in Kusch (2002), knowledge is a social kind, like money, and the status of the possessor of knowledge as regards a certain claim P is simply a nexus of obligations and commitments held by a community of people towards the claim that P.

A disadvantage of Gilbert's account is that it does not explain the exact relationship between individual brain states and group states. The question of whether the group state of belief that P supervenes on individual states is unaddressed whilst the notion of belief is replaced with the notion of commitment, where the status of the latter as a behavioural trait or a mental state is left unexplained. This however need not affect the communitarian account to be presented below, since the purpose of the account is to shift the unit of epistemic analysis as regards scientific knowledge to the community rather than the individual, rather than give a complete mechanism of how and if the group states reduce or relate to individual mental states. The latter endeavour is deemed irrelevant, since the stance followed is that focussing on group behaviour has more explanatory power in analysing the production of scientific knowledge than focussing on individuals' cognitive capacities and faculties[15].

I will close this section by elaborating on the communitarian schema of knowledge acquisition by contrasting it to figure 3.

4.2.2.3 Knowledge acquisition according to communitarian epistemology

As hinted in the conclusion of the previous subsection, knowledge acquisition of ordinary claims is not described accurately by figure 3, which describes only special cases of acquisition of knowledge through perception or solitary inference.[16]

Given that most of our knowledge is acquired through testimony, and given the importance of communication stressed throughout the above sections, a more complete schema of knowledge acquisition would drive a wedge between the time of grasping that a state of affairs obtains and being justified in attributing knowledge to the subject, only attributing knowledge when a community of people is formed. Such an account is given in figure 4, which is copied from Kusch (2002, 72).

My claim is that the schema below does justice to the arguments presented above, as to the constitutive role of community validation in the acquisition of knowledge. Furthermore, it does justice to the idea entertained more formally in the introduction that the epistemic subject is the group and the individual becomes an epistemic subject as regards a

---

[15] As stated in the beginning of this chapter (section 4.2) it may turn out that a certain change in mental or brain state occurs when one "comes to see" or 'grasps' claim P. However, this grasping for me does not qualify as obtaining knowledge, even if one takes knowledge to consist in justified true belief. The reason for this is that, according to the story I have given above, all three aspects of the traditional definition of knowledge are ineliminably social: justification depends of validation by the relevant community, "ultimate" truth also contains a social element, and finally belief has to do with group states, as Gilbert describes.

[16] However there is a case to be made to the conclusion that inference is based on rule-following, and hence, if one accepts the communitarian interpretation of rule-following, then this would not be available to the social isolate.

proposition P only in virtue of there existing a community which accepts P as a knowledge-claim. This is demonstrated in the schema by the events taking place at "time 4", where a community is formed and only then can the two people forming that community, *a* and *b*, be said to know that P.

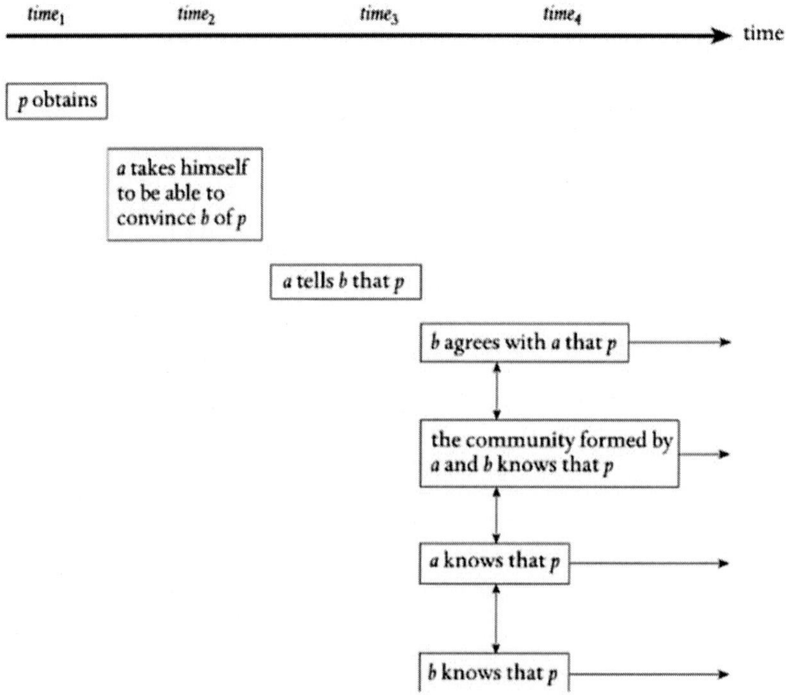

Illustration 4: Knowledge production in *case of constative testimony.* (Copied from Kusch (2002, 72). By kind permission of Oxford University Press.

Finally, it is worth remarking that knowing that P as described in figure 4 and as proposed by communitarian epistemology, consists, as noted above, of a nexus of commitments and entitlements, and hence the content of the concept of knowledge may vary from group to group. This is a social account of knowledge acquisition, and it may be compatible with more formal, traditional definitions of knowledge. A strength of the social account of knowledge that I have given above is that it can explain the claim that there is a vast amount of knowledge "out there" in the world at the same time as considering knowledge to be something that is firmly rooted inside the human brain. To reiterate, according to the communitarian epistemologist as described in this section, the seat of knowledge is not the individual knower or any state inside his mind, but the community that accepts a given claim as a knowledge-claim and acts accordingly. The knowledge-claim just isn't the sort of thing that sits idle in a person's brain or mind, it is rather something that expresses its presence out there in the world through its instantiations by users belonging in concrete epistemic communities. Such uses are subject to the set of commitments and entitlements that the community has as regards granting the status of knowledge to a given claim. Hence the notion of an individual possessing all of the knowledge of a given community becomes meaningless, as, to adopt a more traditional terminology, what the individual's head contains (beliefs) is not justified by anybody. This is because the entitlements and commitments required to qualify something as knowledge are a purely social phenomenon – being rules, they are subject to rule-following considerations which I will elaborate on in the next chapter.

In order to complete the "micro-level"[17] picture of knowledge acquisition in communitarian epistemology, one needs to supplement the picture presented in figure 4 with some at least one more time-slice, regarding the enlargement or not of the newborn epistemic community. At this stage I will focus on the case where claim P is totally "newborn", in the sense that there is no other group that holds that P, something which is which is frequently the case in the production of scientific knowledge[18].

In the first case, and more specifically in the case of scientific communities, what is obtained at the new time-slice is that a person *a* gains a special status and accompanied kudos in virtue of being considered as what I will label in the following section the "closure champion". Person *b*, or more usually, the core set of scientists working on establishing whether P or not-P is the case, gain the extra-celebratory and authoritative status of the possessors of scientific knowledge that P.

It should be noted that the labelling of "extra-celebratory" status as regards claim P concerns the external relations of the core-group, mostly with lay-people as regards such claim, but also to a lesser extent to scientists with other specialisations. As may be discerned from the previous chapters, the status of the possessor of knowledge is a celebratory status that lay-people award according to their own justification and

[17] As opposed to the macro-level picture to be presented below, which is presented in terms of commitments and entitlements.

[18] In the second case, persons *a* and *b* merely enter an already existing epistemic community that has the status of "knowers" of P.

judgement standards, whereas the status of the possessor of scientific knowledge is an extra-celebratory status awarded based on criteria and justification procedures set down by the scientific community at large (such criteria may include reproducibility and peer-review, to name a few candidates). Hence it may be concluded that the keys to justification (based on the definition of knowledge as justified true belief) as regards scientific knowledge are held by the scientific community itself, as opposed to being held by the community of lay-people. This unfortunate situation is one that enhances the feeling of a democratic gap between lay-people and scientific elites, as described in Chapter One.

In what follows, I will attempt to illustrate how the second thesis of communitarian epistemology, that of science designating a social status, applies to scientific practice. I will do this by focussing on the sociological phenomenon of "distance lending enchantment" as applied to the production of scientific knowledge. The discussion on this topic will help shine some light on the "nexus of commitments and entitlements" that I have alluded to above, through a concrete example.

4.3 On "Distance Lends Enchantment"

It is quite a banal observation that the further one is away from an event, both in terms of time and of space, the more distorted the event is presented to them, and it is often the task of historical and sociological scholarship to present a more sophisticated and complicated picture of events than the one that first meets the eye. A second observation is that when one gets closer to historical events, then as more contingencies and hidden motives are uncovered it becomes harder to maintain an unambiguous judgement as regards the event and its protagonists. Conversely, often the more distant an observer is to an event the more certain they are about their assessment of the event.

In the current section I will present evidence that these two observations hold in the domain of science by describing the mechanism of a phenomenon labelled as "distance lends enchantment", first presented by the sociologist of science Harry Collins (1997). I will describe the mechanism by which Collins argues the phenomenon takes place, before going on to examine two cases in the context of decision-making within and based on science. I will argue that one case presents problems for Collins' account, and then go on to present an analysis of the phenomenon based on communitarian epistemology. Finally, I will claim that communitarian epistemology has the resources to offer a more comprehensive explanation for the phenomenon and elaborate on how it does so.

4.3.1 The process of the closure of a scientific controversy

In his paper "Expertise: Between the Scylla of certainty and the new age Charybdis" (1997), Harry Collins divides the interested parties as regards a live scientific issue into four sets: the core- set, the rest of the scientific community, the funders and policy makers, and the general public. He examines the closure of a live scientific disagreement, such as the one on cold fusion. His claim is that two main processes take place, labelled "mitosis" and "crystallisation", focussed in the core-set and the rest of the scientific community[19] respectively. A further long-term process takes place at the level of the funders and policy makers. This last step is less important in the production of knowledge and is associated more with the dissemination and the long-term direction of science. I will now elucidate the phenomenon by elaborating on the composition of the groups along with the three processes taking place during the closure of a scientific controversy.

The core-set comprises of the set of scientists directly involved in the settling of the controversy; they are at the cutting edge of the theoretical and experimental work pertaining to it. These scientists are the absolute authorities as regards the controversy and the first-hand sources for anybody wishing to get involved. Their opposing viewpoints may be triggered both by scientific motives such as personal glory or conviction for a theoretical world-view, and by extra-scientific motivations, such as funding. Collins describes the opposing sides of the debate as non-rigid and not clearly defined prior to closure. This is because they are the only ones aware of all the complicating variables and of all the uncertainty inherent in the experimental process. A further complicating factor is what Collins calls the experimenter's regress. Collins argues that

> Because experiment is a skilful activity, the only way to know whether you have done it right or not is to look at the outcome, but what counts as "the right outcome" is the very cause of the controversy. The theorists too are disagreeing; while each of them individually may be certain that they have a decisive argument, collectively, this settles nothing. Thus, controversy would continue in the core-set for ever if "closure" of controversy depended solely on experiment or theory or their elaborations.(1997, 128–130)

---

[19] To be sure, Collins uses the terminology "scientifically literate commentators" in his schemas, however in his introduction to the paper (1997) he claims that "The rest of the scientific community is made up of scientifically literate commentators".

The process of "mitosis" concerns the resolution of the scientific issue. The issue is typically resolved by a "closure champion", an established scientist who decides the issue by "sticking their necks out" in favour of a view, through a well-written publication consisting of well-argued points. It is such an "audacious" move that polarises the core-set into those critical and those approving of the methods used in establishing the view of the closure champion. After such an event, the majority of the scientists belonging to the core-set swings to the side of the champion, and are less reluctant to produce and publish experimental evidence backing their claim.

The second stage in closure is what Collins calls "crystallization". This occurs in the set of the rest of the scientific community, of which the overwhelming majority sides with the winning view. This group, according to Collins (1997) assigns a higher degree of certainty to the winning view than the core-set members, because the information they are fed about the controversy is often second-hand and comes through polished publications or books which only briefly mention, if at all, the uncertainty[20] and complications accompanying the initial claims. Furthermore, a potentially ex post facto rationalization of the theoretical background of the winning claim is also provided as grounding, with less or no space afforded to alternative theoretical accounts. This is a much more rigid phenomenon, and once a claim crystallizes, it is incorporated into the orthodoxy and is rarely questioned again. Collins goes as far as to characterize the members of this community as zealots on account of their partisan defence of the winning claim (Collins, 1997).

The final stage in the mechanism is that of consolidation, which takes place among the funders and policy-makers. They also side with the winning side, thus reinforcing research by much needed material support and funds. Of course, the door is left open for organizations to invest money in the losing side, on the basis of a Pascal-style wager.

The wider public often only gets to know the details of the winning view and how it was arrived at through idealised accounts praising the triumph of logic and the scientific method. None of the initial uncertainty reaches the public, and it may not even be aware that there was a fierce disagreement in the first place. Those who do not accept the established claim will form an isolated sub-culture, which can and does nonetheless exist and thrive on the margins of science. Such examples are parapsychology,

---

[20] The uncertainty regarding the theoretical and experimental foundations of the claim, and the inherent philosophical limitations of the experimental methods in general should be separated from the familiar error bars and statistical analysis which are present in scientific papers.

creationism and homeopathy.

As Fleck (1981, 113) remarks on a similar note, on what he terms "popular exoteric knowledge",

> Owing to simplification, vividness, and absolute certainty it appears secure, more rounded, and more firmly joined together. It shapes specific public opinion as well as the Weltanschauung and in this form reacts in turn upon the expert.[21]

This, then, is the description that Collins gives of the mechanism of the phenomenon of distance lending enchantment, as regards the closure of a scientific controversy. The characterisation of "distance lending enchantment" articulates the claim that the more removed in social space and time one is from the site of the production of a scientific claim, the more certain they are about their knowledge of the truth or falsity of the given claim.

4.3.2 Decisions based on science and decisions within science

In the above section I have identified a phenomenon taking place within science, and attempted an initial analysis, based on Collins' (1997) idea of the phenomenon of distance lending enchantment. I will now proceed to test Collins' account of closure of scientific issues, by attempting to provide an analysis of decision-making based on science.

Decision-making based on advice of experts from within the scientific community is gaining in importance at the onset of the 21st century, a time when the speed of politics is greater than the speed of science (Collins and Evans 2007, 1). It is also important to keep in mind, as Gregory and Miller remark (2000, 91) that

Public issues tend to arise from new science – from science-in-the-making. In these circumstances, what the public sees is not reliable knowledge, but reliable knowledge in the process of being made, with all the uncertainties and conflicts that inevitably arise while scientific ideas are being turned into reliable knowledge.

I will distinguish between two cases of a live scientific controversy. There are two main criteria for this distinction, one of which is the degree of consensus within the scientific community, and the other whether there is an urgent policy decision to be taken.

The first case is the process when the scientific side of a controversy has more or less been settled, in the sense that there is a consensus among

[21] In this quotation Fleck already leaves open the possibility of feedback loops, something which I will return to below.

the scientists of the core-set as regards the content of the claim, and furthermore there are no pressing policy decisions to be taken. In such a case, the onus is on the funding bodies to decide whether research in the given field of science is worth pursuing and supplementing with material funds. This is especially important, since at the end of the 20th century state funding for scientific research has subsided and has been replaced in part by private commercial sources of funding (Pickstone, 2000, 189).[22] At this stage also, and with a broad consensus among the scientific community on the theoretical issue at hand, other issues such as ethical and philosophical considerations surrounding the field may enter into question. Political and economic decisions may also function as determinants of the fate of research, as is evidenced by the desire of the European Union to establish itself as the leading force in scientific research through the Lisbon Agreement. These considerations are medium- and long-term strategic considerations. In such a case the questions asked are general and regard the usefulness of the area of research itself, rather than the particular scientific controversy. As regards this last question, issues regarding the autonomy of the scientific community as a whole may arise, such as whether non-members of the particular scientific community have the right to scrutinise and impose limits on scientific research. Hence there may be turf wars between disciplines, such as the "Science Wars" in the mid-nineties, which pitted scientists and philosophers of science on one side, and sociologists of science and post-modernist philosophers on the other, regarding the legitimacy to critique the content and practices of science. Another possibility is that of competition among disciplines and scientific establishments for limited funds, as described by Tournon (1993, 95).

In such cases the diffusion of whatever turns out to be scientific knowledge seems to conform to Collins' account at the time after the process of "mitosis" has taken place; there is an "outward" diffusion of knowledge from the core-set, with what is taken as knowledge by one group causing the next group to develop its own knowledge of the subject. Knowledge is transmitted in an authoritative way, with no questioning and with a strict separation of the fields and rank of expertise. This causal chain, as described in Schema One, is one-way, and is in the same direction as that of the distance from the locus of the production of knowledge.

It is worth remarking that in Schema One, at the level of the "scientific

---

[22] See also Nelkin (1984, 24) for some data on funding for biomedical research in academic centres in the US.

community", the rest of the scientifically literate community should be included, and especially other academic disciplines that may take an interest in the issue, but do not interact with it in the same way as the lay public.

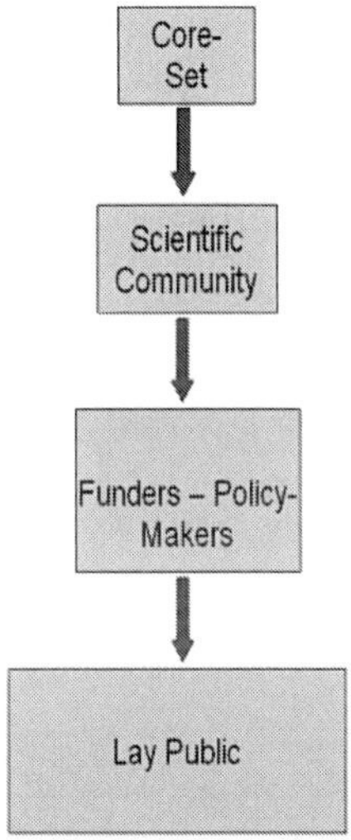

Illustration 5: Schema One: Causal links of knowledge production in a non-u*rgent scientific controversy*

A second more complicated case is the case when the process of mitosis has yet to take place, the "sides" of the controversy are not well-defined, and furthermore the need for a decision is pressing. This is becoming increasingly pervasive as policy and state decision-making is more and more informed by and reliant on science. In such cases where uncertainty reigns supreme, it is easier for factors extraneous to science to influence the knowledge produced. In matters where urgent decisions are required, such as the response to epidemics, the core-set may even be bypassed by outside groups, in an attempt to be seen to be taking precautionary action. Pressure groups consisting largely of lay-people or lobbies attempting to influence policy-makers, and motivated by various interests are formed and may act upon the other groups and in a way solve the scientific issue for them.

An example of such a causal chain is the case of the Brent Spar derelict oil platform. As recounted by Gregory and Miller (2000, 159-160), campaigners from the environmental group Greenpeace boarded and occupied the oil rig, claiming that sinking it in the Atlantic Ocean would endanger marine life, and that furthermore this would set a dangerous precedent for the sinking of other derelict oil rigs. The Greenpeace popular campaign forced Shell to back down and look for other methods of disposal. All along, Shell experts, as well as other experts on environmental issues had insisted that dumping would have been the most environmentally-friendly option. These critics became justified in the eye of the public and in the media only later that year, when the head of Greenpeace admitted that the Greenpeace argument rested on erroneous evidence and estimates. However, by then the decision not to dump at sea was irreversible.

The above incident exposes certain flaws in Collins' account, in the sense that in a case where a decision was urgently needed, the causal chain for the decision taken included feedback loops from one group into the other. In this case, the feedback was from the lay public onto the funders and policy-makers.

A second example regarding a live issue but with no urgent decision depending on it, and exhibiting multiple feedback loops, is one recounted by Jean Tournon (1993, 91-101). The example concerns decisions surrounding the building of the European Synchrotron Radiation Facility. Tournon sketches the way that the rest of the scientific community affected the core-set's desire for a synchrotron radiation facility. As he states,

> a growing number of scientists, running beyond physics into biology, chemistry, geology, and in the applied sciences for calibration and imaging, divided themselves between those who bet for and those who bet against the chances of synchrotron radiation proving a major new technique opening important new applications (1993, 95).

Furthermore, as Tournon illustrates, the core-set of synchrotron radiation enthusiasts devised a strategy using their capacities as both ESF and national government advisers, to lobby for the building of a synchrotron radiation facility. The political decision to commit funds to the establishment of a synchrotron radiation in turn enabled new knowledge pertaining to properties of matter to be produced; hence the funder's decision directly influenced the output of the core-set.

From the above two example cases, I conclude that Schema Two (Illustration 6) is a more accurate representation of the causal chain as regards the decisions taken within science and based on science. Note that the difference in the thickness of the arrows represents the fact that the outward stream of knowledge still carries more authority, and that furthermore the manner in which the causal influence is propagated is qualitatively different in the two flows.

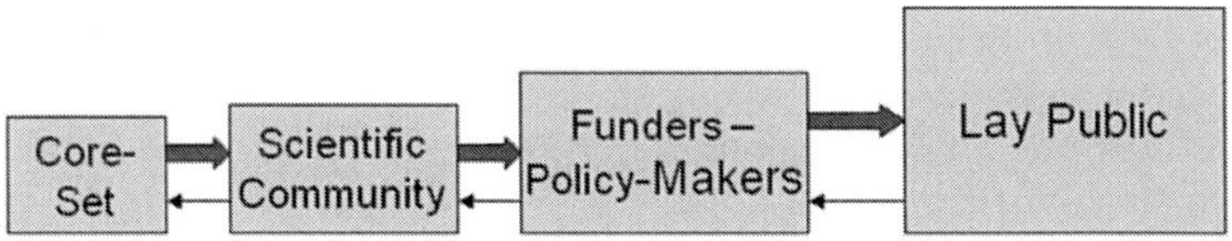

Illustration 6: Schema Two: Causal chain of knowledge production in the *case of an urgent decision*

It is worth remarking that Schema Two is a simplification of actual incidents like the Brent Spar, the European Synchrotron Radiation Facility and the REF, in the sense that it is possible for causal links to exist between any two groups, for example the core-set may be directly influenced by actions of the lay public. Furthermore, it should be noted that the phenomenon of distance lending enchantment is still at work in Schema Two, however it is not sufficient by itself to explain the feedback loops and the influences on decision-making.

I hence conclude that the phenomenon of distance lending enchantment, as presented by Collins, is not sufficient by itself to account for decision-making based in and based on science. I will now explore an analysis grounded on a totally different epistemological basis, which takes into account the phenomenon but nonetheless provides a more complete explanation of decision-making based on science.

4.3.3 An explanation of decision-making in science from the viewpoint of communitarian epistemology

4.3.3.1 Criticism of Collins' account

In the preceding section I highlighted the role of social and other factors external to science (idealistically conceived) in the process of decision-making based on science. I have identified an important slice of decision-making based on science that cannot be dealt with by Collins' account. Following on that analysis, I claim that Collins' account is a rather conservative one. This is because despite identifying the role of extraneous factors in the closing down of scientific controversies, it does not allow for sets other than the core-set to determine the content of the knowledge produced or the decisions made. Collins allows for sociological factors only internal to the core-set to influence the emergence of consensus within that set. Furthermore, it seems that Collins would be happy to subscribe to a "deficit model" of public understanding of science as described in Chapter Two. As Gregory and Miller claim, "the deficit model locates knowledge and expertise solely with the scientists and keeps them at the top of the heap" (2000, 134).

Sociologist Brian Wynne remarks that because of the "top down" character of science communication,

> information that is meaningful in the scientific context in which it originated is likely to arrive in an entirely different lay or public context, in which people are unlikely to accommodate it (should they choose to) without substantial interpretation and adjustment. ([Wynne 1992] cited in [Gregory and Miller 2000, 87])

This fits well with the phenomenon of distance lending enchantment as regards science communication; however it may be insufficient to account for the various pressures that scientists may face when policy-decisions are involved. Furthermore, the deficit model and the top down communication of science fit well only with "settled", often textbook, science, and do less well when live scientific controversies are concerned, when the science is science-in-the-making.

On a related note, communications scholar Robert Logan argues that the diffusion model of science communication

> is essentially a model in which science communication is intended to be purely persuasive (one might say propagandist) ([Logan 1992] cited in Gregory and Miller ([2000, 88]).

That this is the case is illustrated by the vertical structure and the direction of the causal chain in Schema One. However, as the case of the Brent Spar shows, the picture of decision-making is more complicated, and the causal links may exist between any two sets.

I will now attempt to provide an epistemological analysis which I think is more adequate than Collins' account in answering these worries.

4.3.3.2 Communitarian epistemology in a nutshell

Unsurprisingly, this view is that of communitarian epistemology. In the current section I will explain briefly the second main thesis of communitarian epistemology, as laid down in the introduction. I will then explain how it accommodates decision-making as described in Schema Two.

The second thesis of communitarian epistemology is laid down by Kusch (2002) as:

- The term "knowledge" and its cognates (such as "knower") designates a social status

Kusch claims that obtaining the status of "knower" brings with it certain commitments and entitlements. As a social kind then, the attribution of "knower" to an agent, be they a community or an individual, brings with it a certain set of commitments and entitlements. Two further ideas ground the theses of communitarian epistemology. The first is that testimony is generative of knowledge, whilst the second is the idea of "multiple but local consensus" among the members of a social institution. Kusch uses this idea to show how convergence on narrow-bandwidth empirical beliefs is produced, without a master control.

My claim, in continuation of the discussion started in this chapter, is that such consensus-producing interactions take place between all four communities, and that furthermore what the social status of possessing knowledge entails is different for each group. By this I mean that the formation of popular exoteric knowledge and its relationship to esoteric (core-set) knowledge based on the interactions between the four groups that Collins identifies can be modelled using the resources of communitarian epistemology.

I will achieve this by identifying candidates for the status of commitments and entitlements for the four different communities of knowers identified earlier in the analysis.

4.3.3.3 Knowledge as a social status: Commitments and entitlements of different communities

By exposing how each of the four sets identified by Collins bears different commitments and entitlements when in possession of the status of knowledge, I hope to demonstrate that what it means to know a scientific claim depends on the set one is situated as regards that claim.

I will begin by identifying the commitments and entitlements of the core-set. First of all, the core-set is considered to be the ultimate arbiter or authority, either directly or indirectly, as regards granting the extra-celebratory status of "possessor of scientific knowledge of P" (where P is the given scientific claim). This, of course, comes with the commitment of defending their own status as such arbiters in the face of questioning from others who may dispute that claim P be awarded such a status as a "scientific knowledge-claim". The way that claim P may be disputed will be of course through the questioning of the theoretical chain of justification or the soundness of the experimental process, or of whatever other standards are shared[23] between the core-set and other scientific communities.

---

[23] Given that such a process will essentially be an argumentative procedure (see

As regards entitlements, members of this set are entitled either individually or collectively to authoritatively ask for resources for the continuation of their research and to support their case for the importance of their discipline and their own research. However this entitlement stops at the level of the asking and arguing for their case, as the ultimate decision for funding and thus the evaluation may be up to the funders and policy-makers, with the matter of resorting to further expert opinions at their discretion. Another important entitlement is that of being the only group whose questioning of the claim is taken with utmost seriousness, since members of other groups can be dismissed as lacking the credentials and expertise to do so. Thus, in a way, this is the set that "decides" the actual content of the claim.

The wider scientific community are committed to give the claim its air of authority by affirming its legitimacy as regards the procedures that gave rise to its birth, as those are described in the publication. They are also committed to the popularisation of the claim by providing appropriate "simplifications" of the claim for the lay public, such as writing textbooks or popular science books.

A further entitlement that the wider scientific community has is to the judgement of the relevance of the claim to the wider corpus of scientific knowledge, as well as of its importance. This is especially important given that it is to members of this community that funders may appeal to for the evaluation of proposals (as an outsider to the core-set is required, for reasons of objectivity) and allocations of funds and resources.

Finally, the wider scientific community is entitled to use the claim in their work without further questioning, simply be referencing the relevant publications. This entitlement however comes together with the commitment to not treat the given claim with the same critical attitude that they may reserve for their own work[24], but rather with an aura of certainty justified by an appeal to correct procedures and methods. Thus the commitment is that any sociological issues that this group may be aware of would not be propagated to outside groups, thus making it so that the other communities are not aware of these factors and receive a "cleaned-up" version of the events surrounding the production of the claim.

The funders and policy-makers are committed to use the claim as authoritative knowledge in the justification of their policy, and to furthermore cede authority to the two sets described above as regards the

---

Chapter Three), some common rules of the game are sought for in order to start the conversation.

[24] By this I mean work in their area of expertise, in which case they would comprise the core-set.

factual content of the claim. On the other hand, this set actually decides the fate of the research by being the ultimate seat of the decisions on the allocation of funds and other resources for the continuation of the research. Thus they are entitled to "put the research into context", in the sense of judging the actual usefulness of the area of research and the significance of the claim for the wider public. Such an entitlement gives them enormous power over the continuation of the research, something which occasionally leads to turf wars as regards the independence and self-ruling of scientists.

A further entitlement that funders and policy-makers have is to use the claim as a basis for wide-ranging action, such as addressing a social problem or a public health issue. An example of this is the example of the Brent Spar as cited above, where the Prime Minister played an active role in supporting deep-sea dumping (Gregory and Miller, 2000, 159), even though the public campaign resulted in contrary action. Another example of the policy-makers entitlement to action based on a scientific claim is the action taken by the UK government during the BSE scare in both the early and mid-nineties (Gregory and Miller [2000, 172-174] give some details regarding this case).

Finally the lay public's commitments to the claim are, if they accept science as a whole, to cede full authority as regards the factual content of the claim, and furthermore to not question any aspect relating to its production and legitimacy. The lay public, in gaining the status of possessing knowledge of the claim, is also committed to its quasi-blind defence against any detractors wishing to attack science as a whole. Since the wider public is often not aware of the uncertainty surrounding the claim or of any legitimate scientific alternatives to the claim, it is also committed to view the claim as affirmation of the rigour and progress of science in general. In exchange, the lay public is entitled to use the claim as authoritative, incontrovertible and certain knowledge in their discussions, as well as to ground certain value claims and lifestyle preferences on it.

4.3.4 Implications

I hope to have shown from the above analysis of the different entitlements and commitments (based on communitarian epistemology) that what the status of the possession of scientific knowledge entails for different communities of people is simply different depending on how one is related to that claim. I will now relate this to the original discussion of the phenomenon of distance lending enchantment and decision-making based on science.

As regards the phenomenon of distance lending enchantment, it is easy to see that the commitments as regards the certainty of the scientific claim increase as one is removed from the locus of the production of knowledge. Furthermore, since the uses to which the scientific claim is put are different for each community, and given that considerations of uncertainty play a role in the entitlement to the status of possession of knowledge only for the core-set[25] , then it is easy to understand why certainty increases as we move from the core-set outwards.

Thinking of knowledge as a social status entailing certain commitments and entitlements also enables progress on the question of decision-making based on science. The key to understanding how cases such as described by Schema Two can be described through this model lies in the importance of urgency and of decisions to be taken. Given that the decisions to be taken cut across the set boundaries and affect everybody, then it is conceivable that the entitlements and commitments of one group in order to qualify for the status of knowledge possession as regards an issue may clash with those of another. And in such a case, given the discord between the groups, there is scope for any group to influence any other as regards the actual content of the claim to be produced. As Gregory and Miller (2000, 165) remark

> Science-in-the-making puts strains on everyone involved in the process of public understanding of science: on scientists, in knowing what to claim; on journalists, in assessing what is reliable and significant; and on the public and their representatives, in matching the new facts and ideas to what they already know and how they already live, and in deciding what to do.

Gregory and Miller note that the problematic character of such situations is amplified when risks to people are involved, and that in these situations, "working out "the right response" cannot be achieved exclusively from within the realm of science" (2000, 165).

A final note may be made on the model of science communication implied by the account given above. The above account fits well with the idea of communication networks as elaborated by the sociologist Niklas Luhmann (1992). Luhmann conceptualises the boundaries of science as limits of open territories containing communities defining themselves by

[25] Of course, this is a contingent matter and subject to potential change, as the lay public may shift from moving from "knowing a lot of science" to "knowing how science works" and ultimately to "how science really works", see Durant (1993, 131-134).

communication. Hence membership in the community is contingent on sending and receiving of communications. This model may help expose the mechanism by which multiple but local consensus is produced (see Kusch (2002, 155-156)).

4.4 Concluding remarks

In this chapter I have presented the two main theses of communitarian epistemology, and have defended them by way of showing that they help explain two phenomena linked with science, namely the priority of the community over the individual as the epistemic subject, and the sociological phenomenon of distance lending enchantment. In construing the epistemic subject as the community over the individual, I have continued the line of thought elaborated in Chapter Three, that science is social by constitution. In the second part of the chapter I elaborated the idea of the possession of knowledge being a social attribute, defined by a nexus of commitments and entitlements, and proceeded to give an example of such commitments and entitlements for different communities regarding a scientific knowledge claim. In the next chapter I will move on to defend a controversial aspect of communitarian epistemology, that of meaning finitism, before briefly dwelling on a problem regarding the source of normativity.

CHAPTER FIVE

MEANING FINITISM DEFENDED

5.1 Introduction

In this chapter I will defend the position of meaning finitism, as defined by Barnes, Bloor and Henry (BBH) (Barnes, Bloor, and Henry 1996) and used by Kusch in his defence of communitarian epistemology. I will begin by explaining the thesis and address the question of why communitarian epistemology needs it. In section 5.2 I will also elaborate on BBH's defence of meaning finitism and mount a criticism of this strategy of defence; I will present a counterexample that is intended to show why the thesis is untenable. In section 5.3 I will give an example of a domain of words where meaning finitism seems to be a reigning supreme. In section 5.4 I will expand on two domains from within the natural sciences where, in my judgement, meaning finitism is applied. In section 5.5 I will present some thoughts as to which terms should be problematised in semantics, as well as a tentative theory which explains why it is that some terms become susceptible to a meaning finitist analysis whilst others don't; I will be closing with some general remarks on language, before relating back to scientific communities and democracy. Finally, in section 5.6 I will focus on a second controversial thesis needed by the communitarian epistemologist, which has to do with the problem of rule-following and normativity.

It is important to bear in mind where I am starting from and where I am going. I am starting with a prima facie implausible position and a good argument against it, supplemented by a counterexample which makes it seem that meaning finitism is indeed a silly theory of meaning. What I will do is not produce a counterargument for this claim or any argument in favour of meaning finitism. I will provide some grounds for the argument that counterintuitive and implausible as the position may appear, it seems to be, implicitly, the theory at work in a large domain of word meanings. I will then attempt to link it with the particular domain of interest, which is the domain of scientific knowledge.

I will now move on to the first part of the chapter, in which I will lay down the thesis of meaning finitism and explain why it is essential for communitarian epistemology to get off the ground.

5.2 Meaning finitism and communitarian epistemology

In order for the thesis of communitarian epistemology that knowledge always designates a social status to get off the ground, the possibility of personal knowledge (which is knowledge that is not subject to anything social) of word meanings has to be eliminated. It is worth recalling that Kusch (2002, 1) is clear that this thesis of communitarian epistemology is true for all domains of knowledge, rather than true for most domains or most situations.

A potentially thorny domain for communitarian epistemology is knowledge of word meanings. This is a special domain of knowledge which serves to ground all other domains of more complex knowledge (such as propositional knowledge), therefore it is necessary that communitarian epistemology provide an account of it congruent with its two main theses. One way to achieve this is to argue for the thesis that word meaning knowledge is also a social status; or that in any case meanings are social institutions rather than personal states of mind or relations between states of mind and things in the world. The latter option is, of course, the mainstream view of word meaning knowledge.

Now, it may be claimed that meaning finitism as a theory of word meaning is not essential to communitarian epistemology but rather very helpful for it. Kusch is aware of this possibility, however he believes that there is no way of separating the two without falling into incoherence (2002, 174). The additional baggage to communitarian epistemology, such as the views on normativity and objectivity, are necessary for the sake of presenting a complete picture. Kusch claims that the reason that such questions are not addressed by epistemologists is that they are part of the background of the discussion; as such, there is consensus among epistemologists as to what the correct answers are (2002, 173). As communitarian epistemology presents a radical break with traditional epistemology, such questions have to be brought into the foreground and addressed anew.

5.2.1 Meaning finitism described

The features described below are to be found in the definition of meaning finitism, as proposed by Barnes and Bloor, and adopted and elaborated by Kusch in his defence of communitarian epistemology. Kusch summarises the position as the conjunction of five theses, which I will now lay down and briefly describe, based on Kusch's description (202, 206–208):

- Future applications are open-ended (open-endedness thesis) – this means that the correct usage of a term is not fixed in advance, but rather the decision as to whether usage of a term is correct is taken anew every time the term is used. In practice this means that all usage of terms is contestable, not that it is contested every time a word is used.
- No act of classification is ever indefeasibly correct (defeasibility thesis) – this thesis expresses the thought that classification is made on the basis of similarity and analogy rather than strict identity. If the latter was the case, then there would, theoretically, be a single correct classification, which is that of bundling together all things identical (Hesse 1974, ch. 1).
- All acts of classification are revisable (revisability thesis) – this makes explicit something implied by the previous thesis, which is that given that there is no single correct classification, it is possible (and indeed probably quite common) that a classification deemed correct at one time may be deemed incorrect at another time. There may be two main reasons for this, either a "natural drift" of exemplars, or explicit identifiable interests that force a change in classification. In this chapter I will focus more on the latter type of reason.
- Successive applications of a kind term are not independent (independence thesis) – this in a way reveals a Wittgensteinian influence and it may be seen as a consequence of the "meaning is use" slogan. It conveys the idea that past applications of a term have a causal influence on future ones, even if it is not a determining influence (as this would contradict the open-endedness thesis).
- Applications of different kind terms are not independent of each other (entrenchment thesis) – this mirrors an idea introduced by Hesse (1974), that of semantic webs, and is also mirrored in the concept of joint entrenchment as elaborated by Collins (1985, ch. 1). It conveys the ideas that the use of terms in the description of other terms serves to sharpen the meaning of both. Collins gives the example of the terms "emerald" and "green", as an example of joint entrenchment.

Having described the main theses of meaning finitism, I will now move on to address the issue of why it provides support necessary for the thesis that knowledge is always a social status.

5.2.2 Meaning finitism and communitarian epistemology

In order to argue for the thesis that the status "knower of **p**" is solely a social status, one needs to show that nothing other than social factors dictate when it is appropriate to make the attribution of knowledge to a person. Such a thesis seems to fall at the first hurdle as regards propositional knowledge, if knowledge of meanings turns out not to be social or not to be fully social. As stated above, word meaning knowledge occupies a special place in epistemology, as it is the foundation of propositional knowledge. Hence, the following three theses need to be true in order for the communitarian theses to get off the ground:

- Attributing knowledge is irreducibly social.
- Classification of things into kinds is contingent on goals and aims, that is there is no such notion as cutting nature at its joints.
- Given the classification, whether an instance belongs to a certain class of things is up to us – there's no determined fact of the matter about it – that is even granted the above two theses, if there is such a thing as a fact of the matter as to whether, given a classification, things belong to a certain class or not, then the conclusion that knowing meanings is always a social status doesn't follow.

A few observations are in order. First of all, it is worth noticing that the theses presented above begin as fairly harmless for the realist, but augment into a thesis that seems unacceptable. For example, a realist such as Tuomela (2003) would accept the thesis that at least some knowledge attribution is irreducibly social, but does so from a firmly realist perspective. Furthermore, John Searle (1995) would probably accept the first and second theses, at least for the classification of certain terms (what he terms "the assignment of a status function" [1995, ch. 1-2]) whilst still giving a story about how such a move presupposes the existence of a real, mind-independent world.[1]

However, the thesis that word knowledge and its cognates always and without exception designate a social status requires all three theses to be

---

[1] I leave aside the question of the coherence or plausibility of Searle's position, as sketched in Searle (1995).

accepted, including the stronger third one. This is because social statuses have three key features, that they are not mind-independent, that they are not objectively (only at best intersubjectively) correct, and that they can be revoked at any time based on the community's whim. These three features are contained in the third thesis as described above, and are furthermore key features of meaning finitism as laid down above. I will now move on to the next section, where I will give the Barnes-Bloor-Henry defence of meaning finitism, and show why such a defence (or any defence of this type) is doomed to failure. The thesis implicit in the next section is that any attempt to ground meaning finitism on an empirically falsifiable theory of cognition will probably fail, as the domain where the grounding is sought for is itself quite shaky and ever-changing. The proposal will be to look for such an explanation in another domain, perhaps equally "scientific".[2]

5.2.3 Barnes-Bloor-Henry defence of meaning finitism and a criticism

David Bloor and Barry Barnes, the main proponents of meaning finitism and of the more general Strong Programme in the Sociology of Scientific Knowledge, have been insisting for a long time (Barnes 1974; Barnes 1982; Bloor 1983; Bloor 2002; Barnes, Bloor, and Henry 1996) that their views have a strong naturalistic base, something which has made enemies for them both "left" and "right"; they have been attacked for this claim both by natural scientists such as Mermin (1998, 603–623) who are quick to point out that their views on scientific practice are deeply inaccurate and they do not deserve to be labelled as naturalistic and by philosophers insisting that their account is normative rather than naturalistic (Forge 1996 34-40). On the other hand, Bruno Latour (1999, 113-130) seems to find fault with finitism exactly because he thinks it is naturalistic – Latour seems to have rejected naturalism wholesale in an exchange of articles with Bloor (Bloor 1999a; Latour 1999; Bloor 1999b).

Even from their early papers and books on the subject, Barnes and Bloor try to promote meaning finitism as a scientific theory of the

[2] It is worth remarking that what would count as "naturalism" is quite vague: in German, the "extension" of the word "wissenschaft" would probably include domains such as law or sociology, whereas in English it would perhaps be totally silly to call law a "science", let alone grant it the celebratory status of "natural science". Other disciplines such as sociology would fall in a grey area of the "social sciences", with debates raging in philosophy departments as to whether they deserve to be have the same status as the "natural sciences".

acquisition and use of word-meanings, emphasising the role of ostensive definitions in the process. Indeed Kusch does cite (2002, 202) empirical work in the field of cognitive science that lends some support to meaning finitism. Bloor defends the ”scientific nature” of meaning finitism from critics such as Turnbull (1996, 41–44) who claim that adopting a naturalist methodology on a topic precludes the possibility of criticism on that topic. The bone of contention in this particular debate is whether one is allowed to claim, as Barnes, Bloor and Henry do, that their methods are naturalistic and at the same time come out with such a stringent criticism of science. Bloor brushes aside this criticism by claiming that the starting point of a self-understanding exercise on science does not pre-determine its end-point, in other words it is possible to do other things (such as describing and criticising at different times) (Bloor 1996, 52-62). However there are many more ways one may attack a meaning finitism, if its aspiration is to present itself as a scientific theory.

A main criticism is that meaning finitism simply seems implausible as a theory of concept acquisition for children learning their mother tongue. It seems that it is not the case that children learn all words through ostensive definitions, rather they seem to quickly acquire an ability to grasp meanings through the utterance context. Furthermore, it seems that children quickly get “tuned” into a more or less wholesale way of classifying, in the sense that once they acquire the ability to speak, they succeed quite often in using words appropriately, or in any case in a manner that adults approve of. Barnes, Bloor and Henry would have to give a story on how this happens, as one would expect children to make many mistakes in classifying according to similarity as their base from which they draw exemplars, based on family resemblance, would be much more limited than that of the adult speakers of a language.[3]

Barnes, Bloor and Henry do give a story of a brain architecture that suits them, favouring a modular architecture of the mind as proposed by Fodor (1983; 2001), rather than a connectionist one as proposed by Paul (Churchland 1989) and Patricia Churchland (1992). The appeal to an architecture of the mind/brain may turn out to be a red herring for them,

---

[3] A counterexample to this which could perhaps provide support for a meaning finitist account of concept acquisition may be the study of communication among infants who are just learning to speak. In this case, it is plausible that children are 'forging' word meanings local to the context they are at (different babies will have not only different (as well as very limited) arrays of exemplars for each word, but they will also pronounce the few words they know differently). This possibility sounds very promising as setting a defence of meaning finitism, however I will not pursue it further for the time being.

should empirical evidence not go their way. Indeed, it seems that connectionist and dynamic models of the mind have become more mainstream among neuroscientists, hence the preferred architecture linked to meaning finitism may turn out to be rather implausible based on the production of more empirical evidence. Furthermore, such a move renders Turnbull's criticisms more pointed, in the sense that it seems to beg the question to attempt to ground a sharp criticism of science on a scientific theory. If scientific theories are worthy of criticism on their own terms, then surely the basis of the criticism is shaky too. However, pace Turnbull, due to the polyphony of science, such moves seem to be occurring, as for example research in game theory, undercutting theories of rationality and construing reason "purely as an instrument for avoiding inconsistent behaviour" (Binmore 2008, 7), or pluralist theories in the foundations of mathematics supporting the conclusion that terms only make sense within theories or worldviews (Hellman and Bell 2006). Turnbull's criticism probably has bite exactly because BBH's theory is so sweeping, in that if accepted, it denies any possibility of non-relative knowledge – however it seems that at least meaning finitism would have to be considered as absolute (even if descriptive), hence bringing to the forth the charge of the self-refutation of relativism.

In conclusion, it seems that at the moment we are left with a bad argument for an implausible conclusion. In the next section I will attempt to strengthen the opposition to meaning finitism by highlighting a counterexample which will give empirical and philosophical grounds for doubting the theory, in additional to the empirical ones of the previous section.

5.2.4 A counterexample to meaning finitism: Deaf people's communication

As an example of a language domain to test a meaning finitist account for natural language meanings I will choose deaf languages, for two simple reasons: first, they exhibit many of the features of natural languages and secondly, because the objection that I wish to bring out as regards meaning finitism is more clearly illustrated using deaf language as an example.

As regards the first reason, one of course may contest the claim that sign languages are actually languages or that they are structurally the same as oral languages. This objection seems to me to be rather political, relying on a technical definition of language in order to rule out languages that don't fit the purposes of the interlocutor. My only assumption about language at the moment is that it is the main instrument of communication among humans, besides whatever else it may be. A second related objection may be that meanings of terms are simply not the same sorts of things as the assignment of signs to concepts; however, once more, I would respond that the assignment of a linguistic sign (be it oral or signed) to a concept is surely a necessary constitutive part of meaning, and hence this is no threat to my use of sign language as an appropriate example.

Regarding my use of sign language as more illustrative of certain features of natural language, the feature highlighted is that universal signs arise as a result of repeated and spread-out use, out of local dialects. A second feature of sign languages is that they are not phonetic languages, in the sense that things very rarely get spelled out in sign language, but rather signs usually stand for whole words, in the same way as pictographs, and this increases their local and contingent character.

I will now examine which of the five theses of meaning finitism sign languages satisfy, before moving on to the particular meaning finitism thesis that sign languages must definitely not satisfy, if they are to serve their purpose as means of communication. I will then generalise this point as an objection to meaning finitism.

First of all, it is clear that for sign languages, the applications of different words are not independent of each other. This is very obvious in the case of sign language word formation, in that in devising a sign, a speaker consciously combines signs that already exist, thus giving a description of the thing being signified. Hence the entrenchment thesis is satisfied.

Secondly, successive applications of a sign for a term are not independent. It is clear that as the sign gets more widely applied, different users may change it slightly or build on it, plus it may be that the originator of the sign may choose to develop the sign or change it according to the context. Furthermore, it is possible that the same speaker may develop a sign based on previous usage when in conversations with different people. It therefore seems, at least prima facie, that the independence thesis is also satisfied.

Thirdly, it is clear that for sign languages, there's absolutely no reason for classifications to be permanent and non-revisable. Given the isolation of deaf communities from each other, it is entirely possible that when people from two different sign communities meet, classifications would be one of the things that enter the “trading zone” (Galison 1997, ch.9) and are revised into different ones, localised and restricted to the participants in the conversation. This usage may in turn feed back to the larger community by means of a speaker returning to it and introducing the new classification. Hence the revisability thesis also seems to be satisfied.

Fourthly, it is clear that classification is never indefeasibly correct and that it works on the basis of analogy rather than identity. Evidence for this is provided from the fact of sign and classification diversity, in the sense that different communities pick different aspects of objects when coming up with new signs. If there were indefeasibly correct classifications, then one would expect much more uniformity than one does find in fact. One suspects that, given sign diversity, the issue of indefeasible classification would be dismissed as out of hand immediately. Hence the defeasibility thesis seems obvious for sign languages.

So far, so good for meaning finitism and sign languages. However, it seems that the final thesis of meaning finitism, the thesis that future applications are open-ended, is not and cannot be satisfied by any sign language and by extension by any natural language. This conclusion presents a threat to communitarian epistemology, given the key role of the thesis in the communitarian epistemologist's argument. I will now explain more clearly what the threat is.

5.2.4.1 A second objection to meaning finitism

In order to get a good grasp of the meaning finitist account and especially the thesis about the open-endedness of meaning, the meaning finitist has to provide an account of the mechanisms involved by which meanings are negotiated or agreed upon. Lacking this account, there is nothing barring a meaning determinist[4] from claiming that meaning assignments are like function assignments in mathematics, in the sense that once the function is set, hence an operation of the type "let x be y" has taken place, then the dice are thrown for x; it can only be y and nothing other than y.

Furthermore, in order for the meaning finitist to avoid an infinite regress[5] by way of going back to the meanings of the words used for agreement being themselves contestable and in need of agreement upon, then the mechanism of agreements of meaning has to be extra-linguistic.

This problem may be called the genetic problem, in the sense that the meaning finitist has to come up with an account of how first meanings are arrived at, in the absence of any language. This problem also affects Kusch's position of communitarian epistemology, in the sense that Kusch admits (2002, ch.XI) that his account presupposes the existence of social

---

[4] See Kusch (2006, ch.1)

[5] I wish to thank Katie Monk for showing me, perhaps unwittingly, the severity of this problem.

institutions,[6] in order to justify other social institutions, and gives no account of how the first institutions arise.

The genetic problem acquires more bite in the case of communication between deaf people, precisely because the motivating interest of communication is more explicit than, say, the desire to describe the natural world. A further reason why the genetic problem has more bite in deaf people communication is that all sorts of mechanisms such as facial gestures, that, for spoken languages would be considered as extra-linguistic devices, are already part of the codified language, therefore it seems that deaf communicators have no recourse to extra-linguistic devices in order to signal agreement towards the meaning of a word. Deaf communication, hence is a vivid illustration that for any sort of interest to be expressed and communicated and for agreement or contestation of word meanings to take place, then linguistic communication must already be presupposed. I will now explain why such negotiation of meaning does not seem to take place within deaf people communication.

Deaf people, as I have mentioned before, often come up with ad hoc signs in their encounters with other deaf language users. However, once a sign is assigned to a word, then the message that a deaf person needs to give their interlocutor is that all future occurrences of that particular sign do indeed stand for the word first assigned to it. In a way, deaf people need to presuppose the constancy of the link between the signs and the words signified and their meanings, if they are to communicate at all. Their communication would simply not occur if they were required to constantly renegotiate sign assignments. Furthermore, there is no empirical evidence supporting the claim that in fact they do negotiate meanings. Sign assignment, in a "let x (sign) stand for y (word indicator) for all following occurrences in the present conversation" way happens very fast in conversation, in a manner that may even be characterised as automatic. Of course, it may be argued that negotiation of meaning does indeed take place in such encounters, albeit through a mechanism unknown to me. It may be claimed that the pragmatics of the conversation supply this mechanism or that there is a codified way in which it has become embedded in the sign language. Whatever this is, it seems to me that the burden of proof is on the person promoting such a position to show that there is such a mechanism, with empirical observation suggesting, at least prima facie, that no such interaction does nor can take place.

---

[6] Kusch, in line with meaning finitists, considers word meanings to be social institutions.

5.2.5 Interim conclusion

It seems, at the moment, that things are pretty bad for the meaning finitist. Not only have I given an argument that limits their resources in the defence of the position, I have also given a counterexample from an actual language practice that seems to preclude straight away the possibility of any defence of meaning finitism. Hence, the meaning finitist position is left looking more implausible than ever.

In the next section, I will choose not to defuse these arguments, but rather locate a domain of concrete examples where, at least prima facie, meaning finitism seems to be the only theory of meaning that can explain linguistic practice and concept formation. I will then attempt to show, firstly, that this particular domain is not at all marginal and that secondly, terms that have to do with a specific sub-domain of language, that of the language through which science is made and communicated, have been and are susceptible to a meaning finitist analysis. Hence the dialectic strategy will be to show that despite the problems described above for meaning finitism, I am accurately describing linguistic practice when I observe that, at least for certain domains of language, we are all in practice meaning finitists. I will then claim that we are in effect meaning finitists when we engage in argumentation using the language (or languages) of science.

5.3 Protected Geographical Status, standardisation, interests, and halloumi cheese

The domain I will focus on concerns the domain of the definitions of edible goods and what passes as what in the domain of the food industry. There are three main features which make such a domain attractive to the meaning finitist.

First of all, the decision as to whether something falls under a given classification is taken up by law courts, in this case the Protected Geographical Status legal framework of the European Union (Wikipedia 2012a). This means that explicit deliberation is required and is conducted as to whether a given object falls under a given classification, for example as to whether a piece of cheese is indeed halloumi cheese or not. Such decisions are a) not indefeasibly correct, b) can be revised, and c) do not incontrovertibly fix future use of the term. Hence, they satisfy most if not all of the features of the meaning finitist analysis of word meanings.

The second feature which makes this domain attractive to the finitist, is the fact of the plurality of legal frameworks and hence of organisations that "decide" whether instances of things fall under given classifications. Indeed, the practice itself is contested at certain countries and certain trademark law traditions, as for example the United States which is "opposed in general to the protection of geographical designations of origin"(Wikipedia 2012a). Furthermore, even within the traditions that accept the idea of geographical designations of origin, there are different legal frameworks, such as the "Appellation d'origine contrôlée (AOC)" designation used in France, the "Denominazione di origine controllata" (DOC) designation used in Italy, etc. As a consequence, the fact whether something falls within a given classification in this case is a) a purely "institutional fact" (Searle, 1995) and b) it is actually relative, as it may be the case that the very same piece of matter (for example cheese) may be considered as halloumi cheese if one is in the US, but not if one is in the EU.

Finally, interests play a key role in the formation of classifications. As a matter of legal procedure, a case concerning classification is brought up before a court of arbitration in all and only cases in which there is an interest on the part of the plaintiff. Considering the British legal system (admittedly, on the topic of judicial review, but I assume that this is a condition for all cases) the senior Courts Act states that

No application for judicial review shall be made unless the leave of the High Court has been obtained in accordance with rules of court; and the court shall not grant leave to make such an application unless it considers that the applicant has a sufficient interest in the matter to which the application relates (Barnett 2004, 739).

This is important, because a case built on an inherent interest in finding the truth about whether a slab of cheese is indeed halloumi cheese or not would not even be granted a hearing on the basis of lacking what is termed "standing" or "locus standi", which is a term of art meaning "a position from which one may validly make a legal claim or seek to enforce a right or duty" (Garner 2011, 840).Hence, put in a simplified and rather crude way, a judge, when asked to deliberate, by a legal person, as to whether a given slab of white cheese is indeed halloumi or not, will first ask "What's in it for you?" and then think about answering the given question. And what's worse, the answer "I just want to know" will not even qualify as a suitable answer.

5.3.1 A little bit of background – So, what is halloumi?

According to Wikipedia, halloumi or haloumi[7](Greek χαλλούμι, Turkish hellim) (Wikipedia 2012b) is a cheese first made in Cyprus during the Medieval Byzantine period. The same article traces the etymology of the word to Ancient Egyptian and Coptic, which means that either halloumi simply means cheese and the techniques used to make it were simply used all across the region to make any sort of cheese, or that halloumi cheese was made in the region long before the Medieval Byzantine period and in any case it is not of Cypriot origin. Whatever its origins, in modern times halloumi cheese is mostly associated with Cyprus, and it is

currently registered as a protected Cypriot product within the US (since the 1990s, despite an appeal by the Danish Dairy Board, see [U.S. Department of Commerce - Patent and Trademark Office and Trademark Trial and Appeal Board 1999]) but not the EU[8]) (Wikipedia 2012b).

In terms of physical characteristics, halloumi cheese has a distinct squeaky texture, and also has a high melting point, which makes it ideal for grilling. Furthermore, halloumi cheese "is set with rennet and is unusual in that no acid or acid-producing bacterium is used in its preparation" ([O'Connor 1993] cited in [Wikipedia 2012b]). This is quite significant, as, under certain classifications, halloumi would hardly even qualify as a cheese, as one usual definition of cheeses is that they are the products of the fermentation of milk by acid-producing bacteria. However, halloumi is not the only cheese made this way, hence the definition "cheese made without the use of acid or acid-producing bacteria" does not single it out, nor can it be used as a definition for it.

5.3.2 The controversy

One should think that given its physical characteristics, halloumi would be easy to define and there should not be any controversy at all as to what counts as halloumi. However, this is far from being the case; the question is far from settled, what passes as halloumi varies greatly both in

[7] Given the evidence of the pronunciation of the word in other nearby languages, it seems that there needs to be a strong 'l' sound in the middle of the word. The reasons for the double writing in greek may be either because of commercial interests (trademark issues) or because of an effort made in the mid-nineties by a government of the Republic of Cyprus to make things resemble more closely to the standardised greek language as used in greek newspapers and greek government officials. This was also an interest-laden effort, to be sure.

[8] More on the reasons why shortly.

composition, method of production, taste and probably natural properties as well, whilst the story of its definition is laden with intrigue and controversy.

First of all, however, a little bit of Cypriot halloumi history. Halloumi cheese has been made in Cyprus for many many years, with the techniques as to its secrets and its production passing down mainly from mother to daughter, and of course not written in any handbooks. Every self-respecting possessor of more than a few sheep would probably produce their own halloumi cheese at home, with many variations occurring because of the quality of the sheep and their grazing habits.

Furthermore, it was considered a mark of failure for somebody to buy halloumi rather than produce it at home, in the sense that they were too poor to produce it at home. In short, halloumi was not yet seen as a commodity.

When Cyprus changed hands from being a part of the Ottoman Empire to being a part of the British Empire in 1878, and the adoption of the capitalist system in use at the time by the British, halloumi began to be exported more aggressively to other countries, mainly to Cypriots living abroad. At that time the trade was still relatively small-scale, with only some dairies in operation – one of the oldest and bigger dairies and the first to use pasteurization techniques was established in 1939 (Pittas Dairy Industries Ltd 2012).

With independence in 1960, the export of halloumi gained large national significance. Techniques were adopted for the mass production of halloumi, and the first large dairy company, Pittas Dairies, began mass exporting halloumi, targeting not only Cypriots living abroad but also foreigners who weren't acquainted with Cyprus. The mass production of halloumi cheese had three immediate results, which I will mention briefly.

First of all, sanitation controls were introduced. Before the introduction of these controls, what was considered to be halloumi was stored in clay pots at home in its own brine and salt-water. Due to the poverty of houses and the sometimes poor (by our standards) living conditions, the pots would become infested with a certain kind of worm[9]. This worm was considered harmless and caused no concern to the locals, who ate their cheese nonetheless – actually they did not have much choice, being poor. A second feature of sanitation controls was that the milk used was no longer straight from the animal, but it was pasteurised and controlled for fat content. Given that sheep's milk has a higher fat content, home-made good halloumi cheese would probably have a higher fat content than an

[9] Which the locals called "αππηιτούριν" or, literally, the thing that jumps about.

industrially produced one. Hence standardisation contributed in changing the taste and the actual product that was sold as halloumi. It is telling that people in Cyprus in the 1960s and 1970s referred to the cheese produced by Pittas as "Pittas' cheese" rather than halloumi, thus acknowledging the difference. Of course, nowadays no such classification exists, partly because many people up to the age of 40 or 50 grew up linking the word halloumi with the packaged Pittas cheese[10] that says "halloumi" in an iconic fashion on the front (see illustration below) or in any case people now mostly buy their halloumi, which is more often than not, mass produced.

Illustration 7: The iconic Pittas halloumi cheese package

A second change that came about as a result of the standardisation is the change in the actual taste and composition of the product. Halloumi cheese used to be made primarily of sheep's milk, with a bit of goat milk added in if needed. However, due to its higher fat-for-weight ratio, sheep's milk is actually much more expensive than goat's milk and both are much more expensive than cow's milk. As regards Pittas, he quickly fell prey to the temptation of making a lot more profit out of the cheese he was selling, by introducing initially more goat's milk in his mass-produced halloumi and then cow's milk, to the point where most mass-produced halloumi nowadays contains a large majority of cow's milk (anything up to 95%) and very little sheep's milk and goat's milk. Furthermore, this is part of the reason why there is no EU Protected Designation of Origin for halloumi yet (see [Saoulli 2007) – to date, the PDO Application is listed as

[10] On a personal note, the product pictured provided my own exemplar in terms of halloumi taste, shape and concept as I was growing up – even though the shape of home-made halloumi cheese is different, its size a bit larger and its taste different.

"applied",[11] and hence no strict definition of what halloumi is for the EU. In the US, there is a geographical designation stating that the cheese be made in "Cyprus using (the) historic method unique to that country" (U.S. Department of Commerce - Patent and Trademark Office and Trademark Trial and Appeal Board 1999) (this came about as a result of an eight-year court case in which the Danish Dairy Board, Inc., representing the interests of Danish farmers and cheese producers, sought to challenge the registration mark for halloumi so that they could also market and profit from a product that they would also brand as "halloumi"). No minimal percentages are specified, thus leaving the actual taste range and composition of the cheese varying greatly[12].

A final consequence is that due to the more aggressive marketing of the product abroad, many more people became acquainted with the iconic packet of picture one, and have come to define halloumi primarily as that product[13]. Hence, there has grown a world-wide community who uses the word halloumi to designate the product in the cheese section of major supermarkets, which reads "halloumi" on the package and has a particular taste and physical characteristics. Thus halloumi has become "that particular Cypriot" cheese for these people.[14]

It is worth remarking that as long as halloumi sales abroad went well there was absolutely no need to pose the question whether what Pittas sold was halloumi or what really halloumi was. Older people of course still preferred to buy home-made halloumi, since because of modernisation they no longer had any animals at home. The skills involved in the production of halloumi began to die out as common knowledge of the whole population, to the point that it is entirely possible that both my grandparents regarded halloumi-making knowledge as common knowledge, whereas perhaps neither of my parents has the skills or required knowledge to make good halloumi. Nowadays good home-made halloumi is to be found through good connections and personal acquaintance with

[11] See (Republic of Cyprus 2009) and http://ec.europa.eu/agriculture/quality/door/appliedName.html?denominationId=1650)

[12] Of course, as noted above and in footnote 59, setting thresholds on the composition of the cheese and thus making the classification sharper may actually harm the commercial interests of everybody, so probably no court-case is forthcoming.

[13] Compare this with Coca-Cola and Coke in general.

[14] Despite probably the efforts of Lebanese take-away owners, who to this day continue to sell halloumi in pitta bread along with other pastries. Perhaps all they are doing is selling cheese-pies, riding on the wave of publicity that the cypriot trademark had abroad and its effect of disseminating the taste to a larger audience.

small-scale halloumi makers, with halloumi from the village of Pissouri being especially coveted.

Finally, along with the change in tourist target markets and the turn towards artisan products, along with fusion cuisine in gastronomy (see for example [BBC 2012b]), the actual place that halloumi commands and hence the use to which it is put has changed too in recent years. In older years, when halloumi had a higher fat content, and was probably saltier, it used to be eaten raw along with some bread and olives. The two properties of high fat content and a lot of salt made it ideal for herders or farm workers, as the fat gave them energy whilst the high salt content helped to protect against dehydration in a warm climate. Furthermore, mature halloumi cheese is an optional ingredient of flaounes (or pilaounes, as the Turkuphones call them), which used to be an egg-rich pastry made by Grecophone Cypriot women at Easter and by Turcophone[15] women all the time. Besides this, halloumi did not have that many uses. However, this changed when barbecue technology improved in Cyprus and grilled halloumi became a mezze to be eaten along with zivania, the traditional spirit[16] or whiskey or beer, whilst the men would be outside in charge of the barbecue. Two further uses arose in the last quarter of the century among Cypriots; the halloumi in pita bread, a favourite staple food of Cypriot students abroad, and the halloumi with melon or watermelon to be eaten as dinner on the veranda during hot summer months.

As regards tourism, halloumi, now the "national cheese of Cyprus"[17] started accompanying continental breakfasts at hotels, or being served

---

[15] I wish to thank my friend Mustapha Tuncbilek for introducing me to this way of classifying the people of Cyprus, The issue of ethnic origin classification is a delicate issue on the island of Cyprus, because of a lingering political situation. I have found that this classification is more suited to my goals and more coherent with my overall beliefs on the matter.

[16] Zivania, which is made with the distilled product of grape juice, could serve equally well as an example of something ripe for a finitist analysis. The differences in taste between zivania (Cyprus) , grappa, (Italy) and raki (the last word as used by Turks, as opposed to Greeks, since the Greeks call give the name raki to a different drink), and tsipouro (Greece) are left to the discretion of the reader. A further case study would be that of the different nationalities of coffee, since one may be drinking the same drink, but depending on where they are, they will have ordered a Cypriot, Turkish, Greek, Israeli, Arab, Lebanese or Cypriot coffee. This may extent further to Egypt, north Africa, and the Middle East, however I lack the necessary information to support the case.

[17] Where Cyprus in this case stood for the Republic of Cyprus, as opposed to the island of Cyprus or the self-styled Turkish Republic of Northern Cyprus; the latter is mainly shut off for tourism and trade, as a consequence of a trade embargo.

grilled as an appetiser. Finally, in the 21st century, and in the grip of globalisation, halloumi has become available to top chefs (sometimes hired especially for the purpose of increasing halloumi sales abroad), who, along with amateur chefs, have worked wonders in making delicious recipes using it (see for example [Votsis 2012] for an example of the latter).

Finally, with the spread of pasta, halloumi also achieved a role similar to that fulfilled by parmesan cheese, that is grated on top of pasta. Hence it is obvious to see that neither the reference of the word nor the use of the object have remained constant over the years. I will now move back to relating the halloumi example to meaning finitism.

5.3.3 Concluding the halloumi example

I have used the halloumi example as a real-world example of the application of meaning finitism to concept and meaning formation, that is of acquiring knowledge of things in the world and of attaching linguistic signifiers to real-world objects. I hope that the reader has, so far, identified how this example serves the purposes both of the meaning finitist and the communitarian epistemologist. I will, however go briefly through the five theses of meaning finitism, pausing to identify as to how each thesis is satisfied in the case of halloumi. I will also sketch how knowledge of things to do with halloumi fulfil the role that the communitarian epistemologist wants the example to fulfil.

5.3.4 Meaning finitism, communitarian epistemology and halloumi

First of all, it is obvious that the applications of the word "halloumi" when relating it to slabs of white cheese are open-ended and subject to change. An example of this exists with many local variations within Cyprus, as well as for foreign markets with differential legal frameworks. It is still open as to whether that thing will be called halloumi, "pebble cheese[18]" or whether simply people will forget about it and it will disappear. It is also open whether what is now called home-made halloumi will keep being called that, as tastes change and given that the EU may at any point decide to outlaw it for failing sanitary conditions of preparation. Hence the halloumi example satisfies the first thesis of the meaning finitist, that of the open-endedness of terms.

[18] A variation of halloumi made (before 1974) in the village of Akanthou.

Secondly, it is obvious that no act of classification, both as regarding the classification of a slab of white cheese as halloumi and of the class of all halloumi as cheese, or the whole of a certain kind of halloumi (for example that of cow's milk) is indefeasibly correct and unchangeable. In fact, halloumi is now regarded as cheese by many people whereas in the future it may not be so, and furthermore, both the samples of what passes as halloumi would probably not have been considered so by the locals some years ago, and certain cases of what passed as halloumi in the past (mainly that made fully of sheep's milk) would probably not count as halloumi by different tribunals or different linguistic and cheese-connoisseur communities.

Thirdly, it follows from the above point that classifications are revisable, pending mainly on the financial interests of the key players inside the dairy business, but also pending on general rules of trade, the general economic system that dictates these rules and also pending changes in jurisprudence. For example, it is conceivable that in the future laws will settle disputes such as what counts as halloumi through molecular analysis of its composition and its physical properties.

Fourthly, successive applications of the kind term "halloumi" are not independent of each other. This is demonstrated from the fact that applications of the term "halloumi", for example by hungry tourists that became acquainted with the thing-in-the-world that came in the package shown in Picture 1, will affect the overall production of halloumi and in the long term may serve to restrict the application of the term only to that product, thus making the term apply to a particular cheese that will have a smaller range of taste and composition, as the production techniques and the taste requirement of it becomes standardised.

Fifthly, the applications of different kind terms are not independent. In this example, the applications of the terms "cheese", "halloumi", "hellim", "pebble cheese", "Akanthou cheese" are not independent from each other, rather the definition of one reinforces and co-determines the definition of the other.

I will now briefly elaborate on the role that interests play in the argument as to whether something is halloumi or not, in line with the support that the communitarian epistemologist requires of the meaning finitist.

To repeat, the communitarian epistemologist demands that even given a classification, there is no fact of the matter as to whether something belongs to a certain class of things or not. Or, equivalently, the communitarian epistemologist requires that such a fact be an institutional fact (in Searle's terminology) rather than a "brute fact" (Searle, 1995, 27).

The above distinction serves to highlight the mind- and community-dependence of such facts, claiming that even granted the existence of facts, their nature[19] is purely social and thus depending on the people. This reflects, in a way, the Protagorean statement which is regarded as the first statement of a relativist position, that "Man is the measure of all things: of things that are, that they are; of things that are not, that they are not." ([Protagoras, DK80b1] found in [Diels 1903])

Claiming that there either is no (classically defined, such as "[absolutely] true proposition") fact of the matter or that interests are constitutive of any sort of facts regarding meanings, and that furthermore the decisions as to whether an instance of something falls under a given classification, is the key premise that the communitarian epistemologist takes from the meaning finitist and, furthermore, the most distasteful one to their detractors. In a previous section, I also hope to have given a good argument showing how this agreement cannot be taking place within the context of a conversation between speakers.

However, through the halloumi example, I hope to have at least given some air of credibility to this thesis. It is obvious that there is no classically defined – that is mind-independent and community-independent – fact of the matter as to whether a given slab of a white cheese-looking substance possessing a certain set of physical and chemical properties, is actually halloumi or not. Indeed, it is the case that a slab of halloumi bought in the United States would probably not be recognised as such in an EU country, say Sweden. This is because the halloumi sold in the United States is required to consist of sheep's and/or goat's milk, whereas the halloumi sold in the EU consists of a mixture of sheep's, goat's and cow's milk. Hence it may be a fact of the matter for average (as regards their halloumi-related knowledge) people in Sweden that a package with the inscription "halloumi" bought from the US is likely to have been involved in a mix-up or has gone bad or is the victim of false advertising, to name just one example, as their exemplar of halloumi is radically different in many of its relevant (taste, primarily, and also grilling properties) physical and chemical properties. Hence a Swedish person and an American person may get involved in a faultless disagreement not only regarding the taste[20] of halloumi and whether it tastes better than

[19] The question as to the nature of facts, that is what facts actually are, has been one of the most tortured questions in 20th century philosophy, perhaps because of the linguistic turn in philosophy. However, I will attempt to steer clear of such a question.

[20] MacFarlane (e.g. his [2007] has been heavily involved in promoting a relativist semantics and defending a relativism about truth on matters of taste. However, my

gorgonzola, but whether a single object is actually halloumi or not.

Furthermore, it gets even worse for the halloumi-realist. What the average (as regards their halloumi-related knowledge) Swede or American will come to regard as halloumi is inextricably linked to the interests of institutions or groups of people such as the Cyprus Cows' Milk Producers Organisation, the Cyprus Standards Setting Agency, the Cyprus Cheese Producers Associations, the Danish Milk Producers representatives, the EU legal advisers, Lebanese, Syrian chefs and restaurants the judges of any cases brought upon them, and perhaps many others that make a living out of halloumi. And that is only to discuss the most explicit and tangible interest, which is the economic interest. Other more abstract interests such as class interests, macro-social interests, political interests and national interests have yet to be introduced into the picture.

So, to conclude this section, here is where we are at: The detractor of the communitarian epistemology position has given a good argument against meaning finitism as a general theory of meaning, accompanied with an example that is designed to show the folly of meaning finitism. The meaning finitist (and the communitarian epistemologist standing behind him) has not answered any of the arguments given nor the example, but has rather given an example of a very big domain of things in the world whose meaning assignment seems to be more congenial to meaning finitism than to any other position. Given that I will ultimately come out in favour of meaning finitism, it is my task to advance the argument in their favour. So my next two moves will be to show that, first, irrespective of any general aspirations, there is good evidence to suggest that at the domain of entities and processes that play a crucial role in science, meaning assignment seems to conform to meaning finitism and, furthermore, to give an argument as to why it is sufficient for the meaning finitist to give an account of domains of meaning assignments in which their position is the most accurate one in describing practice (in true naturalistic fashion), and that they need not worry about the genetic problem or arguments using aspects or variants of it. I will, thus, set forth with the first task I have set.

5.4 Science and meaning finitism

My strategy, in order to show that meaning finitism and the relevant aspects that it donates to communitarian epistemology fit the domain of

---

thesis is much more ambitious than that.

scientific knowledge[21], will be to identify terms used to describe objects of study of different natural sciences and see if I can give a meaning finitist analysis of these terms. I will attempt to show that there is good evidence to suggest that at least some concepts essential to and paradigmatic of at least some natural sciences are contested and their meanings stipulated among scientists, as opposed to comprising immutable natural kinds consisting of any sort of unchanging "essence". My strategy will be the following: I will begin with a rather "soft" example, that of psychiatry and move into an example much closer to the heart of the arch-rationalist, that of protein identification. The rationale behind this choice of examples is the following: I will first begin with an example that is more akin to Strong Programme examples in the sense that a clear sociological explanation is obvious as regards the interests that drive classification change. This will serve as a heuristic stepping stone that will allow me to proceed into a domain which is closer to hard-core "natural science" but in which such interests are less obviously discernible. Hence, in a way, the first example serves to prepare the ground for an attack to the heart of the arch-rationalist's claims against meaning finitism; an attack which is stronger exactly because it shows that meaning finitism correctly describes changes in the meaning of "natural kind" terms, even in the obvious absence of non-epistemic interests on the part of the actors – although it is worth noticing that the avowed desire for instrumental success and growth of knowledge may be considered as an interest in itself. In doing this, I fully acknowledge that my first example may face the objection that it is irrelevant to the argument at hand, since it targets what may by many be regarded as a non-(natural) scientific domain. In answering this objection, I would point out that the difference between the two cases is one of degree rather than one of kind, in that it interests are present in both of them; the difference being that in the mental illnesses case such interests are more easily discernible than in the proteins case.

5.4.1 Mental illnesses and the DSM

5.4.1.1 Psychiatry, natural kinds and family resemblance

The first example that I will consider as exemplary of meaning finitism at work within the broad domain of natural science is from the field of

---

[21] I leave open the question as to whether there is something particular about scientific knowledge that makes it susceptible to meaning finitism, or whether my conclusions extend to all knowledge, as my topic concerns scientific knowledge.

psychiatry. Psychiatry is obviously not the most paradigmatic of natural sciences, with its scientific status being hotly contested. However, there is a small philosophical literature on psychiatry as a science (e.g. [Hempel 1965; Szasz 1960; Laudan 1983]), whereas the field of philosophy of psychiatry has been rapidly expanding in the last twenty years (see for example [Hacking [1995; 1999],; Cooper [2007],; Thagard [2008]). Nothing much hangs on this, since the object of study is the classification and disease "entities" that make up the Diagnostic and Statistical Manual of Mental Disorders, more widely known as the DSM.

As regards mental illnesses, there is a debate as to whether they should be thought of as natural kinds or not. According to Zachar (2000), they should not; he rejects the claim that if a biological (genetic) basis is found for a mental illness then that is sufficient to ground the claim that this illness does indeed constitute a natural kind. Instead, Zachar labels mental illnesses as "practical kinds", stressing that such kinds "cannot be fully defined with respect to their inherent properties". (2000, 172)

On the other hand, Cooper (2004) claims to refute what she terms "the historical argument", as well as Hacking's arguments against mental illnesses being instances of natural kinds, and argues that at least some mental illnesses, such as those that have a genetic base, are natural kinds terms. She gives the example of Huntington's Chorea, which is caused by a genetic aberration, concluding that

> (p)lausibly, Huntington's Chorea is a natural kind of mental disorder; in all cases an identical underlying property, the defective gene, produces characteristic symptoms. (Cooper 2004, 8)

Other mental illnesses she labels as "partial kinds"; in such cases, according to Cooper's account, "cases of the disease will be similar to each other in many, but not all, respects." (2004, 8)

However, she does allow for the possibility that some mental illnesses may turn out to be neither natural nor partial kinds, in the sense that different instances of the illness will not have any interesting similarities.

I will now move on to some remarks on the history of mental illness classifications before examining how classifications can best be explained by a meaning finitist analysis.

5.4.1.2 History of mental illness classifications

The endeavour of looking at the historical origins and development of disease classifications and more specifically mental illness classifications will pave the way for a look at the interests at stake, and ultimately into whether mental illness terms are susceptible to a meaning finitist analysis.

Hence, I will now begin with a brief look into the history of disease classification schemes. I will look at different classification schemes currently in use as regards to their history and interconnections.

There are currently at least three classifications of mental illness (or disorders, I will use these two terms interchangeably) that are currently widely used: DSM-IV-TR, ICD-10 and CCMD-2-R. These are produced and sponsored by the American Psychiatric Association, the World Health Organization and the Chinese Society of Psychiatry respectively. I will begin with the history of the International Classification, the ICD-10.

Classifications of disease first appeared in the 18th century, with François Bossier de Lacroix (1706-1777) (better known as Sauvages), a contemporary of Linnaeus, credited with producing the first one entitled Nosologia methodica (World Health Organization, 2012). The first classification adopted internationally was the "Bertillon Classification of Causes of Death", named after the chairman of the committee of the International Statistical Institute charged with the task of producing an internationally-accepted classification. In the 20th century the Fifth International Conference for the Revision of the International List of Causes of Death (which took place in 1938) recognised the need for a

> corresponding list of diseases to meet the statistical requirements of widely differing organizations, such as health insurance organizations, hospitals, military medical services, health administrations, and similar bodies. (2012, 3))

Thus standardization of diseases and causes of death became a priority. From 1948 the responsibility for the preparation of the International Statistical Classification of Diseases, Injuries, and Causes of Death was passed on to the World Health Organization. International interest in the Classification grew enormously by 1975, when the International Conference for the Ninth Revision of the International Classification of Diseases met in order to prepare the Ninth Revision and by the 10th revision in 1990, the ICD became the main standard for the classification of diseases.

As regards mental health disorders, the first extensive consultation process took place in the context of the Eighth Revision of the ICD (ICD-8 (World Health Organization 1993), whereas since the 1970s and the revolution in psychiatry happening then, the ICD and the DSM classifications began to converge greatly and research made for revisions on one of the two systems had repercussions for the other. Agreement between classifications became a key interest and priority for committees working on revisions of the classification systems. The third system considered here, the CCMD, also converges to a large extent with the other two, however there are still differences in classification, changes which also affect the diagnostic procedures and outcomes.

In this section I hope to have shown that disease classifications are mostly historical enterprises and have a clear historical root in the 18th century "natural history" endeavour of making classifications. As Pickstone remarks,

Natural history was a common term in the seventeenth century (…). It was the register of facts, the compilation of what was in the world. It contrasted with natural philosophy, which was the account of *causes,* a matter of explanation rather than inventory. (Pickstone 2000, 60)

Furthermore, I have shown that as interests changed in the 19th and 20th century and as divergences emerged between classification systems with different origins, convergence and standardisation among systems became a key interest. As regards specifically mental illness, the revolution in the 1970s played a key role both in the redefinition of mental illness, and as regards convergence among different systems.

Before I review some of the non-epistemic factors that played a role in the 1970s revolution in psychiatry, I will expand a bit on some of the divergences between the systems as well as between diagnoses within the same system.

5.4.1.3 Divergences among and within classification systems

There are two ways in which the systems diverge. The first one is on classes of mental disorders which are considered culturally-specific, something which is more evident in comparison between the CCMD-2 and ICD-10 or DSM-III. Hence, as Lee remarks,

> Thus, the particular additions (e.g. travelling psychosis, qigong induced mental disorders), deletions (e.g. somatoform disorders, pathological gambling, a number of personality and sexual disorders), retentions (e.g. unipolar mania, neurosis, hysteria, homosexuality), and variations (e.g. depressive neurosis, neurasthenia) reveal not only the changing notions of illness but also the shifting social realities in contemporary China (Lee 1996, 421).

Another explanation for such discrepancies is given by Zheng et al., where they suggest that

> such discrepancies may result either from frequent changes of the diagnostic terms in the West such as the phenomenon of neurasthenia, or

> from creating a new disorder entity such as eastern gymnastic exercises-induced mental disorders in the CCMD-2 (Zheng et al. 1994, 441).

A second way in which systems may diverge is on the consistency of diagnoses for disorders which are present in all of the compared systems. In this mode of divergence it is even possible for systems to disagree with earlier versions of themselves, as the symptoms judged as defining of the disorder may change from one version to the next. Furthermore such lack of consistency may be observed even within the same system, in the sense that it is not the case that a given patient will always be given the same diagnosis, even by doctors working within the same classification. In an interesting cross-system study on substance use disorders, comparing DSM-III-R, DSM-IV and ICD-10 some of the major findings were the following:

> (1) Cross-system agreement for Dependence was generally high, especially between DSM-III-R and DSM-IV.
> (2) Cross-system agreement was lower for DSM-III-R and DSM-IV Abuse and very low for DSM-IV Abuse and ICD-10 Harmful Use.
> (3) Agreement varied across drug categories with lowest DSM-III-R/DSM-IV agreement for alcohol abuse and DSM-IV/ICD-10 agreement for marijuana use disorders. (Rounsaville et al. 1993, 337)

A threat more serious than the misidentification of substance abuse versus harmful use is the one demonstrated by an experiment published in *Science* in 1973, conducted by the psychologist David Rosenhan (1973). Rosenhan managed to fool mental institutions into misdiagnosing healthy patients as suffering from mental health disorders, as well as fooling them into misdiagnosing real patients as imposters.

Rosenhan's experiment took place in 1973, at a time when DSM-III was being prepared and when psychiatry was undergoing a major shift.[22] According to Mayes and Horwitz (2005, 249), at that time mental illnesses "were transformed from broad, etiologically defined entities that were continuous with normality to symptom-based, categorical diseases."

In fact, Rosenhan received an angry rebuke from the chairman of the task force charged with the third revision of the DSM, Robert Spitzer.

[22] The most recent reincarnation of this experiment was run by the BBC's 'Horizon' science programme in 2008, where out of a group consisting of five subjects with previous diagnoses and five subjects with no previous diagnoses, three mental health experts correctly diagnosed two out of the five subjects with previous diagnoses, misdiagnosed one, and diagnosed two out of the five 'healthy' subjects as suffering from mental disorders. (see [Spotlight Radio 2012])

Spitzer claimed in his article, referring to Rosenhan's experiment: "This rather unremarkable finding is not relevant to the real problems of the reliability and validity of psychiatric diagnosis and only serves to obscure them" (Spitzer 1975, 442).

This experiment, along with a campaign by gay rights activists, hastened the change in perception as regards the nature of mental disorders. I will now briefly expand into other social factors that also played a role in this sea-change that took place in the perception of the nature of mental disorders.

5.4.1.4 Social factors and the change in perceptions of mental disorders

The dominant narrative regarding the change that took place in the field of psychiatry in the 1970s and was expressed in the transition from DSM-II to DSM-III is that this change was the result of the expansion of scientific knowledge on the origins of mental illness, as well as the medicalization of a number of behaviours which were regarded as more or less normal in the years before. However, Mayes and Horwitz (2005) reject these two claims, arguing instead that the expansion of psychiatric knowledge had already taken place within the paradigm of dynamic psychiatry, which was based on life history rather than symptoms and was the predecessor paradigm which got replaced in the 1970s. The two authors go on to claim that

> Likewise, although the DSM-III did greatly increase the number of specific diagnoses, it did not increase the number of behaviors that psychiatry laid claim to treat. Psychoanalysis had already medicalized a vast number of problems over the course of the twentieth century. (Mayes and Horwitz 2005, 251).

Instead, the two authors blame a range of social and economic factors that led to the paradigm change and that played a role in the switch to symptoms-based diagnoses from the focus on lifestyle issues and life histories. I will review some of them here.

The first factor that led to the change was a general crisis of legitimacy of psychiatry in the eyes both of academia and the wider public. This crisis is most reflected in the writings of the psychiatrist and academic Thomas Szasz and French philosopher Michel Foucault, as well as in popular films such as "One flew over the cuckoo's nest". Mayes and Horwitz further cite as reasons for the disenchantment over psychiatry the marginal status of psychiatry within the medical profession, the reluctance of insurance companies to reimburse patients for "talk-therapy", the need for deinstitutionalization and the treatment of formerly institutionalized seriously mentally ill patients within the community, the increasing influence of medication treatments and the growing treatment of patients by non-physicians such as clinical psychologists, counsellors and social workers (Mayes and Horwitz 2005).

As regards the change itself, Mayes and Horwitz claim that what it reflected was a need for the standardization of diagnoses demanded mostly by insurance companies. Cooper (2004) surveys a number of ways in which insurance companies pressurised the committees drafting the DSM into including new diagnoses[23] and to massage diagnoses by either undermining or overplaying their severity.

Another issue, which Mayes and Horwitz claim had a political role in the DSM, was the question of whether homosexuality was to be classified as a disorder. This forced psychiatrists to address the issue of what would be classified as a disorder, something which led to symptoms-based definitions, largely as a result of political wrangles among psychiatrists polarised into two camps. The first consisted in those who favoured the Kraepelian view of mental illness as defined by "well-defined, specific criteria as the basis for diagnostic decisions"(Mayes and Horwitz, 259) with Spitzer, the chairman of the committee, adopting the controversial slogan "mental disorders are a subset of medical disorders" (Mayes and Horwitz, 260); the second group consisted of those who were more favourable towards Freudian analyses of mental illness which did not emphasise, biological, biochemical, nutritional and neurological aspects of mental illness.

Finally, both Mayes and Horwitz, as well as Cooper, stress the role of the pharmaceutical companies who had an abundance of newly-discovered substances but needed well-defined symptom-based disorders in order to market these substances as cures for them. This push resulted in the phenomenal success of Valium and Prozac in the 1980s and 1990s.

Having laid down the social factors and interests that guided the change in the classification of mental disorders, I will now move on to examine whether the theses of meaning finitism apply to them.

---

[23] An example of this is Post Traumatic Stress Disorder , which was the result of the push to make the Veteran's Administration pay for the treatment of many Vietnam war veterans.

5.4.1.5 Mental illnesses and meaning finitism

From the analysis of mental illnesses and their classifications given above, there are already some hints that there is fertile ground for a meaning finitist analysis. However, I will now go back to the theses of meaning finitism and check whether they apply in this case.

I will begin with the open-endedness thesis, which states that the meaning of a mental illness term is not fixed in advance, but can be negotiated anew every time the term is used. The most obvious contestation and negotiation of the terms takes place during revisions of the diagnostic manuals. In such cases special committees deliberate on many aspects surrounding both what symptoms define which illness, but furthermore it may be decided that groups of symptoms formerly thought of as defining a single illness may not do so any more or may single out a different disease or condition. Hence successive editions of the diagnostic manuals may be seen as exemplars of the open-endedness thesis at work.

The second thesis of meaning finitism is the defeasibility thesis as regards classifications. Given the diversity among different classification systems, as well as the revisions made to them, it seems that this thesis is also satisfied. This thesis is also demonstrated by the variations in diagnoses made by psychiatrists, in the sense that it is entirely possible that two psychiatrists may differ in the diagnosis of the same patient.

The revisability thesis also seems to be prima facie satisfied by various examples in the classifications, most notably by the case of neurasthenia, which has been retained in the ICD-10 and the CCMD-2, whilst rejected by DSM-IV, as well as homosexuality, which was deemed a disorder in DSM-II but was removed from its next revision as well as from subsequent editions of the DSM.

The independence thesis also seems to be satisfied, as there is evidence of continuity of classification terms in successive editions and revisions of the manuals. The most characteristic case may again be neurasthenia, with the term having its origins in the mid-nineteenth century. The term has undergone several redefinitions, however it maintains its historical links with earlier definitions based on psychoanalysis.

Finally, entrenchment also seems to be a well-established phenomenon within classifications of mental disorders, with terms such as substance abuse and substance dependence reinforcing the definition of each other, as well as the definition of the "condition" of homosexuality reinforcing the family of gender-related disorders and the categories of bipolar disorder and schizophrenia reinforcing each other.

Given the above as regards the social and historical factors as well as the semantic ones favouring a meaning finitist analysis of mental illness terms, I conclude that one would be justified giving a meaning finitist analysis of such terms. However, given that there is dispute about whether these terms constitute natural kinds, I will now move into a different domain where there is far less dispute on this question. This will also allow me to examine the role that the assumption of natural kinds plays in shaping the categorisations and meanings of terms studied by the natural sciences.

5.4.2 Protein classification

The example of mental illness terms seems to be a good example of meaning finitism applying to natural science terms, complete with identifiable interests and other social factors which seem to motivate the classifications. To be sure, the proponents of the Strong Programme in the Sociology of scientific knowledge, of which meaning finitism is a key tenet, do not seek to eliminate the pursuit of instrumentally adequate knowledge as a motivation of scientists in favour of the pursuit of social interests, but rather they want to view these two pursuits as two sides of the same coin (Kemp 2003, 311-338).

Hence in the second example of meaning finitism, that I will now present, I will not enquire into specific social interests and their role. Instead, the strength of this example is that it gives a meaning finitist analysis of things which would be regarded as paradigmatic natural kinds – proteins are, after all, macromolecules, hence the thesis of microstructuralism, the view that chemical kinds can be individuated solely in terms of their chemical microstructure (Bird and Tobin 2010) ought to apply to protein molecules as much as it would to chemical element molecules. Hendry (2006) seems to acknowledge some of the problems associated with applying microstructuralism to molecular compounds, citing the heterogeneity of molecular structures at the molecular level as well as the vagueness associated with the notion of sameness of molecular structure. There are many problems, such as limitations to computational power, the sheer diversity and number of different protein molecule types, and other difficulties related to the production of primary data related to protein structure that all contribute to divergences and inconsistencies arising between different software programmes that make up protein classifications.

5.4.2.1 Protein classifications: databases and techniques of production of primary data

Proteins form a huge family of biomolecules and play a very important role in all cell activities. Most proteins are made up by the combination of sequences of twenty standard amino acids, the coding for which is to be found in the genetic code.

There is a multitude of specialised databases which provide information on specialised topics regarding groups of proteins. To give an indicative number, the UniProtKB/Swiss-Prot database cites in its statistics 131 other databases, which it cross-references (http://web.expasy.org/docs/relnotes/relstat.html). UniProtKB cites in the same page that on the 14th December 2011 it contained 533657 sequence entries. However the number of separate protein molecules may be lower as there are sometimes redundancies in data (that is double counting) or sequences that form only parts of protein molecules. It is significant however to note that 70.1% (http://web.expasy.org/docs/relnotes/relstat.html) of the entries are inferred from homology, that is their existence is inferred by the presence of clear orthologs in other closely-related species. Orthologs are genes and their products found in different organisms that are vertically descended from a common ancestor. It is worth mentioning here that the judgement whether two structures are orthologs is a similarity judgement as regards their sequence – there exist many different algorithms as well as computer programmes detecting orthologs and separating them from other pertinent evolutionary relations among protein molecules, such as paralog relations. Such algorithms (e.g. reciprocal smallest distance algorithm [Wall, Fraser, and Hirsh 2003]) or software programmes and associated databases (e.g. INPARANOID, [Remm, Storm, and Sonnhammer 2001) are, however, ultimately are validated through comparisons with "trusted" datasets of orthologs, where "trusted" in the case of INPARANOID is for example

> a dataset of 5500 worm and mammalian proteins in which orthologs had been assigned by manual analysis of phylogenetic trees. In that study, orthologs were assigned if a majority of nine different phylogenetic methods supported the orthology. (Remm, Storm, and Sonnhammer 2001, 1047)

Phylogenetic methods in turn usually consist of the alignment of amino-acid sequences, either manually using human judgement or using automated methods. In order for such a method to be used, a fairly complete sequence of the amino-acids consisting the protein is needed, hence some sort of input from UniProtKB is needed. UniProtKB may in turn incorporate information about homologs in its own database, which may be used as input in databases such as the CATH Protein Structure Classification and the SCOP Structural Classification of Proteins. In return, UniProtKB incorporates links to the classifications (if they exist) in its own entries. This is a nice example of how the different software resources actually depend on each other creating feedback loops.

As opposed to UniProtKB, the Protein Data Bank (PDB) contains data regarding the structure of proteins. PDB contains 78628 structures (on Tuesday 17th January 2012), as opposed to 73507 verified existing proteins in UniProt. It should be noted however that there is no one-to-one match between entries in the PDB and in the UniProtKB, despite increasing efforts in ensuring there is cross-referencing. The Protein Data Bank focusses on structural data for proteins, as opposed to sequencing, which is the focus of the UniProtKB. PDB primary data consists of data on the positions of different atoms within protein molecules, obtained mainly through X-ray crystallography or NMR spectroscopy. PDB structural data are classified in a number of different ways by more than 100 databases. Out of these, I will use the example of SCOP and CATH, as the two purport to have the same aim and share many assumptions and are, furthermore,

> widely used as gold standards to benchmark novel protein structure comparison methods as well as to train machine learning approaches for protein structure classification and prediction. The two hierarchies result from different protocols which may result in differing classifications of the same protein (Csaba, Birzele, and Zimmer 2009, 23).

According to the same authors, their discrepancies are of utmost importance given that they impede the development of fully automatic structure classification methods.

This difference in classification among the two methods forms the core of my claim that classifications of proteins are a good example of meaning finitism at work. Hence I will now elaborate on the differences between SCOP and CATH, before moving on to testing whether all the theses of meaning finitism are satisfied by protein terms.

5.4.2.2 SCOP and CATH

SCOP and CATH, as cited above, are the largest databases of structural relationships among proteins. Structural relationships among different protein molecules are used in many domains within bio-informatics. Hadley and Jones (1999) cite some of these uses as the analysis of protein structure, the extraction of homologous structures, the testing of prediction methods and the calculation of numbers of folds and families. Further applications are in the production of fold recognition templates – such fold templates, aimed at the identification of distant superfamily members, depend on highly accurate groupings of homologous sequences, such as would be expected in homologous families in SCOP and CATH. They may be used in protein structure modelling (Hadley and Jones 1999). Csaba et al. assign further important roles to the two classifications, claiming that

> In the last years, SCOP and CATH have been used to address various questions in structural biology and are further employed as training and gold-standard databases making them invaluable resources in structural bio-informatics. They have been used to study the interplay of protein structure and protein sequence evolution or to explore the connection between alternative splicing and protein structure evolution. (Csaba, Birzele, and Zimmer 2009, 24)

Furthermore, the authors claim that the two classifications are very important for machine learning, as they provide reference and template datasets for benchmarking and evaluation of their performance in automatic protein structure classification and protein structure prediction.

The two databases share some features, whilst having differences in others. They are both hierarchical, but differ in that CATH is semi-automatic, with less human judgement intervention, whereas SCOP is entirely manually curated. Their classification schema also differs: while SCOP uses the categories of Class>Fold>Superfamily>Family>Protein Domain>Species, CATH uses the main categories of Class>Architecture> Topology>Homologous Superfamily, with further smaller family category groupings in a tree structure. The categories of the respective databases do not strictly correspond to each other, however there is some equivalence between the superfamily in SCOP and the homologous superfamily in CATH, as well as the folds in SCOP and architecture in CATH, and both classifications have as their "unit" the protein domain, however their respective definitions of domains are different, something which yields differences and some degree of incommensurability in higher levels. According to Hadley and Jones, whose paper is an early attempt in 1999 to sketch and account for the differences in classification between SCOP and CATH,

> The definition of domain obviously leaves some room for interpretation, and, in some cases, dividing a protein along a possible structural-domain boundary may in fact divide one active-site region into two or more nonfunctional segments. (Hadley and Jones 1999, 1104)

The two authors continue further, explicitly couching their comment in terms of differing epistemic interests. They describe examples where one classification may present as a single domain a protein site which the other classification may split in two or more domains:

> The decision is in many respects a philosophical one: whereas those interested in the biochemical aspects of protein structure may see the structure as a complete functional unit, others with interests in the dynamics of protein folding may argue that the functional unit can be separated into smaller, commonly occurring structural domains. (1999, 1105))

Other differences in the methods employed by the two classifications are attributed, according to a later attempt to indicate the way towards consensus between the two databases, to the different initial purposes of the two classifications:

> CATH is directed more toward defining and classifying the geometric similarities between proteins in order to support and guide structure determination studies and structural genomics. SCOP was originally intended to illuminate evolutionary relationships between proteins, but its curators recognize its utility in more general geometric classification. (Day et al. 2003, 2153)

Before moving on to examine whether the above evidence is sufficient to ground meaning finitism for protein classifications, it is worth noting that there are attempts to build a "standard of truth" for machine learning consisting of the subset of data, in which both classifications agree, (Csaba, Birzele, and Zimmer 2009), and that there are constant attempts to make the two classifications consistent, or to explain away the differences and in any case downplay the inconsistencies. The papers cited above ([Day et al. 2003], [Hadley and Jones 1999], [Csaba, Birzele, and Zimmer 2009]) are instances of such attempts and provide detailed statistics as regards agreement and disagreement among the databases.

5.4.2.3 Protein classifications and meaning finitism

It is now time to address the question whether protein classifications satisfy the theses of meaning finitism.

The first thesis of meaning finitism is the open-endedness thesis, which is the claim that past usage of a term does not determine future applications of the term. This thesis seems to apply to the usage both of protein-terms and of higher protein family terms. In cases of the first kind, the identification of protein samples is quite tricky given the existence of closely-related protein structures and sequences as well as of isoforms of proteins – furthermore, such identification usually relies partly on database searching, which has its own limitations, such as incomplete or corrupted data. Secondly, proteins often have more than one name, with UniProtKB often yielding a "recommended name" as well as "alternative names". Presumably, the realisation that two protein names such as "egg-lysin" and "sperm-lysin" designate the same natural kind term is far from trivial, whereas for many entries in the primary databases such as UniProtKB, there are varying degrees of confidence as regards their existence, with a typology of five types of evidence, titled "evidence at protein level", "evidence at transcript level", "evidence from homology", "predicted" "and uncertain", each associated with its own criteria. Furthermore, the databases are "interactive" in that users' input serves to alter the classifications, something which means that a protein sample identified as **X** at time **t** is not surely to be identified again as **X** in a future instance time **t'**. In cases of families of proteins, it is obvious that there are at least some cases where classification terms may be common in both SCOP and CATH, however the proteins consisting the grouping may be different, hence every time such a protein family name is used for a given protein sample, there is at least the choice to be made whether it refers to the SCOP or the CATH grouping. A more sweeping contestation of protein terms takes place during revisions of the databases themselves, a process which is more or less continuous. Such revisions, as well as the revision of the five types of evidence for each protein entry, are revealing of a process of contestation as regards protein terms.

The second thesis of meaning finitism is the defeasibility thesis, which is the thesis that no act of classification is ever indefeasibly correct. This is an interesting thesis, in the sense that whereas microstructuralism is quite plausible, in practice scientists, at least as regards structural classifications (such as CATH and SCOP), are quite happy to acknowledge the subjective character of the classifications, and are happy to settle for "accurate and reliable" information for all biologists. Hence, Hadley and Jones note that

> To a large extent, the three databases (CATH, SCOP and a third one called FSSP) agree on classifications; certainly no one method is distinctly superior. Most of the differences and discrepancies that exist result from the unique guidelines by which structures are classified within each database. Biologists should note that there are no fixed principles of protein structure classification, and each method relies on independently devised rules. (Hadley and Jones 1999, 1111)

They further advise biologists that

> At present, using these databases in conjunction with human judgement and biological knowledge should be sufficient for providing accurate and reliable structural information to all biologists. Whether a consensus database, devised by extracting undisputed protein classifications from SCOP, CATH and FSSP, would improve the development of accurate threading templates is currently being assessed. (1999, 1111)

Hence, it is recognised from the outset that such classifications are subjective, to a large extent, and aspirations of "correctness" are left aside.

The third thesis is the revisability thesis, which is the thesis that is perhaps the one most obviously satisfied: all papers refer to specific versions of the databases that are current at the time of the research for the paper, and all databases mentioned here feature in their websites the detailed logs of revisions.

The fourth thesis is the independence thesis that states that future applications of a term are not independent of past ones. This is displayed by the continuity of terms even when revisions take place – the revisions made are causally linked to the previous version of the classification, as well as to "new evidence".

Finally, the fifth thesis of meaning finitism is the entrenchment thesis, which states that usage of different classification terms in conjunction sharpens the meaning of both terms. An example of this thesis at work is the use of entries P19448, Q01380, and Q01383 (to take just a few) as well as the term "egg-lysin" in UniProt. The entries have some differences among themselves (mainly in the organism where they are found as well as their length) but altogether (as well as with other entries which are not cited here) define the "extension" (to use a term from traditional semantics) of the term "egg-lysin". At the same time the labelling of the coded entries as "egg-lysin" helps define their functional and structural properties.

In the light of the above, I am led to the conclusion that there is, at least prima facie, a good case supporting the claim that meaning finitism is at work in the classification and use of protein-related terms.

5.5 Meaning finitism, contestable words, meanings and language

In this section, I will attempt to support the restricted conclusion that some kind terms are susceptible to a meaning finitist analysis, and that these terms have, for good or for bad, actually played a role in scientific enquiry. Of course, the second part of the claim I have already achieved for some key terms in the previous section. In this section, I will attempt to gesture towards a theory of natural kinds as based on family resemblance, as well as add some speculative thoughts grounding my wider philosophical position.

5.5.1 Meaning finitism and natural kind terms

Given the above discussion, I will take the opportunity to introduce a theory of meaning that seems to naturally fit my project of defending meaning finitism. This theory is of course not my own invention, but has its root in Wittgenstein's *Philosophical Investigations* (1953) and his remarks on family resemblance. Wittgenstein's work was taken further by the work in the 1970s of Mary Hesse (1974) on semantic nets, as well as the prototype theory of concept structure and acquisition promoted by the experimental work of Rosch and Mervis (1975). What Rosch and Mervis show is that the psychological processes of word categorisation often consist of similarity judgements of each new instance of a term with prototypical exemplars of the category in question, rather than the satisfaction of necessary and sufficient conditions for membership of the particular term in the category in question. Rosch and Mervis's work adds weight to all five of the theses of meaning finitism, whilst at the same time dealing a significant blow to the significance of the above discussion, on whether mental illness terms constitute natural kinds or not. Hence, with mental illness as more generally with all categories, the question of natural kinds or of "cutting nature at its joints" becomes a meaningless question to begin with. However, as Margolis and Laurence (2011) remark, the prototype theory of concepts has its problems, the chief two being compositionality, that is the fact that complex categories (such as "PET FISH") display emergent properties that none of the constituents possess; and furthermore that certain complex categories seem to lack any sort of "prototype" which serves as the blueprint for category membership. At present I will not address any further these worries, but I will conclude this section by reiterating the remark that the question of natural kinds is not significant, nor even perhaps a genuine question within a meaning finitist framework, due to the use of a different theory of meaning.

I will now attempt to speculate a bit on why it is the case that so much of analytic philosophy, and more specifically philosophy of science, has ignored the meaning finitism present in many domains of language usage. I will also look at why meaning finitism and the social nature of science are to be seen as preconditions of cultivating science as conducive to a more democratic citizenship.

5.5.2 Language use and communication

My brief analysis of some linguistic issues that have plagued philosophy in the 20th century begins by redefining the purpose of language.

My claim is that the primary function of language is communication among similar beings rather than denotation of things in the world and description of this world for its own sake. In these terms, there is nothing special about the tools used for communication by humans, that is by human language and the concepts it expresses.

Ignoring this idea leads into pernicious results once it is inferred that somehow humans have privileged epistemic access to how things really are and how they really work, as opposed to animals. An extension of this idea is that there is something specific in the languages (or minds or brains or whatever) of humans that allows them to have a direct interaction with a mind-independent reality. Hence humans have aspirations both for their language and for their activities which go way beyond merely communicating among themselves. The aspiration becomes the one to "cut nature at its joints" or to "describe the world as it is".

What a Protagorean relativism such as the one I am advocating in the present thesis does is to bring back some humbleness into humans as epistemic and semantic subjects. Protagorean relativism restores the idea that we are our own yardstick, that any sort of activity that we engage in is first and foremost a human activity, that it is judged and perceived as such by other humans. This does not discount the possibility that other beings, such as cows or gods, can have their own existence, language, thought and activity, and that their ways of being are relative to their own condition. Furthermore, there may forms of knowledge other than those than can be achieved by the methods of certain human activities, with science as the exemplar.

5.5.3 Going back to scientific communities and democracy

In this section I go back to the relationship between science and democracy, in the light of the remarks given in this chapter so far.

I will provide certain steps towards an antidote to the condition of absolutism. As a first step, we could recognise the contingent – but at the same time constitutive of our thought (the work of Lev Vygotski, which is gaining prominence in education, is a good example of this idea) –

character of language and its use, along with the sheer variety of linguistic practice and use. A second step would be to recognise the relative character of the knowledge claims that are expressed in language. Finally, a third step would be to use the work done in the previous steps in order to recognise how and what we can aspire to is the consensus of the community and at best the consensus of the community of all humans, even as regards scientific knowledge.

The above steps bring to light a tacit assumption of the overall argument of this thesis, which is that if scientific knowledge is treated primarily as relative and as the product of democratic deliberation and consensus among communities of peers then this conclusion, if cultivated among lay-people, will reinforce democratic citizenship. The assumption implicit in this view is that the open-endedness (in the sense of having no fixed answer) of knowledge-claims as well as the status of being in possession of knowledge being a social status are presupposed for a democracy-conducive science. This claim, strong as it may sound, is justified in the following way: Science, as shown more persuasively in Chapter Four, is a social argumentative activity taking place within innumerable fora which are the relevant scientific sub-communities; closure in controversies is achieved when consensus is achieved in the relevant sub-community, however this consensus is often temporary and may be questioned by the same or another sub-community at a future time. Hence, the knowledge-claim produced is at least fallible and open to renegotiation, and furthermore it is only treated as knowledge when it is recognised as such by a specific community at a specific time, based on a specific perception of justification standards employed. What the protagorean relativism that I use as background to this thesis donates to this world-view is simply that this is all there is to scientific knowledge, that the ultimate yardstick of scientific knowledge is a specific community consisting of experts and stakeholders, as well as other epistemic communities, that grant a given scientific claim the celebratory status of knowledge. This, however, is sufficient to ground democracy as well. Democratic methods are used to deliberate on open-ended questions in which some action has to be endorsed by the whole of society or the specified stakeholders in order to be enacted. Democratic deliberation has no place where absolute questions which may be settled once and for all, through the use of a specific method, are involved. Hence, as opposed to the received view, that it is nature that decides scientific questions, I advance the claim that nature is one of the inputs, with the most important input being agreement among a community of human researchers.

In the next section I reinforce the idea that all we have is the community by looking into Kusch's argument about the community being the only source of normativity. By putting forth the notion that the community is what decides what counts as "right" as opposed to merely "seeming right". I also make a nod to the very reasons that appeal is made to democratic decision-making. This is so because when something is put to a vote, it is implicitly recognised that the ultimate arbiter for the question being posed is the community, by way of majority.

Furthermore, this section will serve as closure to the discussion as regards auxiliary theses to the main theses of communitarian epistemology, paving the way to discuss communitarian science in the next chapter.

5.6 Community and normativity

Going back to Kusch's articulation of the theses of communitarian epistemology, I will review the second main thesis of communitarian epistemology, which is the claim that the social status, that is knowledge possession, is typically granted to communities rather than individuals.

In his defence of knowledge as a social status, Kusch stresses that to possess knowledge of a claim **p** is to have certain entitlements and commitments as regards that claim. He adopts the central terms "entitlement" and "commitment" from Brandom (1994), and states explicitly that "It seems right to say that sharing knowledge with others amounts to sharing entitlements and commitments with them".(Kusch 2002, 59)

Examples of entitlements include the entitlement to draw on the claim, as a standard for what others ought to think in particular conversations, and the entitlement to draw on the claim as a premiss in arguments. This sentiment is mirrored in Tuomela's and Balzer's account of "we-thoughts"; However, Tuomela and Balzer shy away from labelling their account as one of knowledge rather than belief:

> The we-mode involves "we-thoughts" in the sense of group members being collectively committed to seeing to it that what is collectively accepted is made correctly assertable or regarded as correctly assertable by group members for the group members. The group members, accordingly, are committed to employing the accepted sentences as premises in their relevant inferences and as grounds for action when functioning as group-members. (Tuomela and Balzer 1998, 148)

A final commitment may be added, that of extending the knowledge-claim to other people, thus enlarging the epistemic community.

The motivation for this thesis stems from the consideration of the rational constraint on beliefs, that is the considerations about what it is that makes us consider certain beliefs as rational and others as irrational. The distinction "rational versus irrational beliefs" is a distinction intended by Kusch to place under the same heading the distinction between justified true and non-justified true beliefs and the distinction between reliably caused true and not reliably caused true beliefs (2002, 87). The answer given by Kusch on the question of the rational constraint on empirical beliefs is that the grounds for this distinction are to be found not within the individual's system of belief, nor in the outside world but rather at the level of the community.

Kusch's dialectic is to begin from purely individualistic epistemological theories of obtaining knowledge, and by criticising each position for its individualistic elements, to slowly move to a full-blown communitarianism. In order to establish communitarian epistemology as the only epistemological theory that can explain empirical knowledge, Kusch's central questions are second- order questions about rationality, such as "Why is it that the dichotomy 'rational versus irrational' can be applied to empirical beliefs?" and "Are there constraints on possible empirical beliefs in virtue of which some empirical beliefs are rational and others are irrational?".

Kusch criticises foundationalism and coherentism on the grounds that these two positions both seek to find the grounds for rational constraint of beliefs in the individual mind, but individual-psychological facts or worldly facts about the mind on their own are not normative, hence they cannot create an obligation or duty on the rational belief creation process. The crucial question that needs to be answered of any normative account of rational belief, according to Kusch is, following Wittgenstein, whether the account has the resources to draw the distinction between "is right" and "seems right". According to Kusch, such a distinction needs a standard of rightness independent of individual judgements and both coherentism and foundationalism, with their very individualist conception of the epistemic subject, simply lack the resources to provide such a standard.

Kusch identifies certain elements that hint towards the importance of the epistemic community in direct realism and reliabilism, however he criticises these two positions as still failing to make sense of the rational constraint upon belief, in the sense that they too take the role of social factors too lightly.

In moving closer towards communitarian epistemology, the positions of Lehrer's (see for example his [1987]) social epistemology and Davidson's (see for example his [1991]) "interpretationalism" (the term is used by Kusch) are discussed, but are also ultimately rejected on the grounds that they still fail to account for the normative constraints upon belief. Hence, Kusch settles on the proposal of communitarian epistemology by a process of elimination.

It is worth noticing that the focus on the community is dictated by the stance that Kusch takes on normativity. He gives an interpretation based on his reading of Wittgenstein, arguing that the most we can get on applying the distinction between "is right" and "seems right" is community agreement. In a sense, what is right is what seems right to the community, hence social isolates cannot be said to have the ability to make this distinction and therefore fail to follow rules. Hence, it is only community adherence to the rules governing language use and belief formation that guarantees the existence of a standard of rightness, in terms of the use of words and of the judgement that certain beliefs are rational and others not. Kusch ascribes to a strong version of the community membership thesis, defending a thesis he calls strongest present-tense community thesis, which states that "An individual is able to follow a rule only if the individual is currently a participating member of a group in which the very same rule is followed by other members"(2002, 181).

This thesis, is meant to be contrasted with accounts of rule-following that allow for the possibility of solitary rule-followers, as well as with weaker community theses. Such arguments usually revolve around the issue of private language, with intuitions diverging on the possibility of such a language. Those who claim that private languages and solitary (that is essentially rather than accidentally private) rule-following is possible advance the position that mental or physical facts about the individual are sufficient to explain rule-following. I will briefly address three families of arguments in favour of this position, together with the communitarian response to them, following Kusch's responses to each challenge.

The first family of arguments groups together two purported arguments that Kusch labels the metaphysical and the internalist arguments, attributed to Colin McGinn (1984). The metaphysical argument claims that it is not inconceivable that God could create a solitary rule-follower, whilst the internalist argument claims that it is only mental facts internal to the individual that determine whether they are following a rule or not. The answer given to these two arguments is along the same lines, that they are not really arguments at all rather than restatements of the positions, in that they are question-begging, and hence do not really advance the discussion. A further argument related to McGinn is the invention argument, which is the argument that if rule-following requires community membership then rules could not have a first follower. Since rules clearly have a first follower, which is the inventor of the rule, it is not the case that rule-following requires community membership. The answer given to this is to break the link between the inventor of a rule and its first follower. This can be seen more easily in the case of community-involving rules such as the rules of team games. The individualist could use the distinction between social and individual rules and insist that in the case of the latter their argument still goes through. In answer to this the communitarian could distinguish between rules concerning individuals (such as solving a Rubik's cube) and rules created and followed by social isolates. They could then question whether necessarily solitary rule-following, such as a social isolate discovering a Rubik's cube and by trial and error learning how to solve it, really is rule-following behaviour after all. In order to do this they would claim that in the latter case, the distinction between "seems right" and "is right" would not apply.

A stronger individualist family of arguments is what Kusch terms the parity arguments, attributed mainly to McDowell (1984) and Gerrans (1998). Proponents of such arguments claim that there is symmetry between the individual rule-follower and the community; and that they stand or fail together on the question of the distinction between "seems right" and "is right". A similar proposal is to replace the community with a community of different time-slices of the same person. Hence, the claim would be that the important aspects of intersubjectivity involved in rule-following could be replaced by intrasubjectivity. To these claims, the communitarian could answer that an important aspect of rule-following is the sanctioning by others and the process of negotiation and convergence towards the right application of the rule[24], something that cannot take place in the same manner within the same individual. Of course, the issue remains that what "seems right" from within the community "is right" for that community. However the communitarian answers this objection by claiming that all judgements are perspectival and that any judgement regarding community error will be based on the exemplars and consensus within another community or of another time-slice of the same community.

Finally, the similarity arguments provide what Kusch judges as the stronger family supporting the claim of solitary rule-followers. Such arguments, whose general line is that we can easily imagine congenital isolates displaying behaviour identical to that of rule-followers, give

[24] It is worth remarking that in line with meaning finitism, Kusch supports the claim that the application of rules contribute to the stipulation of their content, hence the content of rules is not fixed but changes all the time based on the applications.

examples of various thought experiments supporting this claim (Gillett 1995; Baker and Hacker 1984; Blackburn 1984). A first communitarian answer to such a line of examples is to reply that in order for rule-following to take place, the actors themselves ought to have the concept of correctness and incorrectness, but the examples given merely impose our own concepts of correctness on the actors – a counterexample given is one of lower animals which display behaviour similar to what we would label rule-following but however we would be disinclined to grace them with the label of rule-followers. Finally, a second line of response to the given examples is that the purported reasoning patterns are far too complicated and in a way presuppose standards of correctness rather than derive them on the basis of such reasoning.

As applied to epistemology then, the position of Kusch on rule-following would dictate the thesis that an individual can be said to be justified in having a given belief if and only if they are a member of a community which accepts the given belief as justified. However, in accord with meaning finitism, individual instances of interactions between members of the community result in the judgement that a certain belief is justified and also modify the communal array of examples of what is regarded as justified belief. Hence, there is a two-way fit between the individual and the community.

5.7 Concluding remarks

In this chapter I undertook a defence of perhaps the most controversial component of communitarian epistemology, that of meaning finitism. After explaining why this component is a necessary one for communitarian epistemology and finding current defences wanting, I proceeded in a defence of my own, first concentrating on every-day language and then turning back to science. After offering some speculative thoughts on language, science and protagorean relativism and after linking the discussion to the discussion of democracy laid down in earlier chapters, I examined a further element of communitarian epistemology, that of normativity, by presenting the main individualist charges at communitarian conceptions of normativity and elaborating Kusch's answers to them. To conclude the current work, I will now sketch the contours of a "communitarian" field of scientific endeavour, that of sustainability science.

CHAPTER SIX

AN EXAMPLE OF A "COMMUNITARIAN" SCIENTIFIC DISCIPLINE

6.1 Introduction – Recap

The topic of the current work is the notion that communitarian epistemology as regards scientific knowledge prevents the formation of scientific elites within otherwise egalitarian participatory democracies. These elites consist of scientific experts that have an input in policy-making through their input into scientific aspects of societal problems.

After elaborating on this relationship in Chapter One, I proceeded in Chapter Two to look into data regarding lay perceptions and knowledge of the processes of scientific knowledge production. After enquiring into data on the topics of transparency and trust in science and scientists, I proceeded to show how popular science writings help propagate elitism as regards scientists and science as an institution.

In the third Chapter, I returned to the relationship between science and democracy, by identifying argumentation as a common defining feature, and locating the school as the institution where lay-people may be schooled to regard science and democracy as closely linked. In the same chapter I began to give an indication of the wider philosophical implications of the epistemology that I regard as conducive to science being an institution that supports rather than subverts participatory democracy.

In the fourth and fifth chapters I proceeded into a defence of communitarian epistemology, in the first instance as accurately descriptive of scientific knowledge production (Chapter Four) and in a second instance as philosophically sound (Chapter Five), by defending two of its most controversial aspects, those of meaning finitism and of the strong community thesis on normativity.

In this final chapter I return to the link between communitarian epistemology and democracy, by giving a concrete example of a domain of scientific practice and knowledge production that can serve as a blueprint for communitarian science for democratic policy-making. This domain is the domain of sustainability science, a relatively young field of research which has nonetheless risen to prominence in the beginnings of the 21st century.

6.2 Sustainability Science

Sustainability science is a relative newcomer in the field of institutionalised science, with the US National Academy of Sciences (NAS) granting it a space of its own alongside traditional scientific disciplines only as late as 2005, with its birth, at least in as regards the NAS, displaying features characteristic of knowledge being born out of agreement in a communitarian fashion: a committee study proposing the declaration was conducted, followed by extended discussion, ending with the Editorial Board of the PNAS (Proceedings of the National Academy of Sciences) declaring Sustainability Science as a scientific discipline (Clark 2007, 1737). From the outset, sustainability science presents a break with more "traditional" or "paradigmatic" natural sciences such as physics and chemistry, in at least two crucial aspects: it is explicitly problem- and goal-oriented and furthermore it is avowedly value-driven. Furthermore, the problems, goals and values driving such research are more often than not external to sustainability science, and often external to natural science as a whole, being of a political or social nature. This is in stark contrast to traditional conceptions of "fundamental" or "blue skies" research in the natural sciences as curiosity-driven and ostensibly value-free[1]. I will briefly elaborate the first two of these aspects, before devoting some more space to the role of the key value of sustainability.

As noted above, sustainability science is often oriented towards solving problems external to science, something which, according to Clark (2007, 1737) brings it into the company of fields such as agricultural science and health science. However, such a characterisation contrasts sustainability science to both basic and applied science and research. This is because the traditional dichotomy between basic (or fundamental) research versus applied research is usually framed as one between satisfying purely epistemic goals (such as human curiosity and thirst for unifying explanations) on the one hand, and producing products, usually marketable ones, on the other. The products of sustainability science projects are solutions to

[1] For a discussion of these characteristics that are traditionally attributed to paradigmatic natural science, and my own stance on these issues, see sections 1.1-1.2.2

particular problems. Such solutions are not theoretical, as with fundamental research, but consist of solid and practical policy recommendations[2]. Sustainability science research has been labelled as "use-inspired", as opposed to curiosity-driven fundamental science or inventive technology (Kates 2010, 27). The problem-oriented character of sustainability science has a further consequence, that of its epistemological and methodological core being empty, or rather "crowded and heterogeneous" (Nowotny, Scott, and Gibbons 2001, chap. 12). The latter description, as a charge against a scientific discipline, is usually taken to imply that it has no distinct epistemology and methodology, but rather opportunistically uses the methods of other fields of enquiry, which are thus viewed as "non-empty". By this I mean that there is no set of methodologies nor a theoretical body of knowledge that characteristically constitutes or defines "the science of sustainability". Quite on the contrary sustainability science is inherently interdisciplinary and freely utilises knowledge and methods specific to the problem at hand.

A second characteristic of sustainability science is that it is goal-driven in a way that paradigmatic natural sciences are not. Sustainability science research is avowedly teleological, with its goals and aims of sustainable development being explicitly non-epistemic. In this it is in contrast with those conceptions of science that only accept epistemic goals as the primary ones for scientific research, with other goals such as proposing policy regarded as secondary or in any case not defining of scientific research.

Finally, sustainability science is avowedly and unashamedly value-laden, driven by the value of sustainability. In this respect, Weber's notion of value-relevance may be of some application here. According to Parsons, (1965, 173) Weber introduces a distinction between value freedom (Wertfreiheit) versus value-relevance (Wertbeziehung) both of which he regards as crucial for his methodology. Roughly, according to value-freedom, sciences have their own embedded values which take precedence over other concurrent social values for the scientific practitioner. Furthermore, according to value-relevance, there is an element of relativity in research work especially in the social sciences, introduced by the value-system of the researcher themselves, alongside those of their wider society. However, this does not result in a loss of objectivity, as objectivity regards the procedure and empirical investigation of research propositions.

The admission of a strong role for values is significant, given that

[2] I will return to the topic of the relation between policy and sustainability science further down in this chapter.

values, more so than concepts such as "facts", are much more readily susceptible to a meaning finitist analysis. Definitions of values differ according to different linguistic and cultural communities, as well as within the same community of users through the passage of time. Finally, the non-separation of the role of facts and values as the motivation of research within sustainability science, give a specifically local and relative character to the knowledge produced. Having recognised, at least prima facie, the role of values in sustainability science, in the next section I will attempt to elucidate the value of sustainability and its constitutive role in sustainability science research.

6.2.1 Sustainability: an ill-defined value driving a scientific discipline

The concept of sustainability or sustainable development is one with a clear political history and genealogy, however its definitions may run into the hundreds (Kemp and Martens 2007, 7). A widely accepted "baseline" definition of the term is the one linked to its emergence in the world policy scene, through what has become known as the "Brundtland Commission". The Brundtland Commission, more formally known as the World Commission on Environment and Development (WCED), was initiated by the United Nations General Assembly in 1982, and published its report, chaired by then prime-minister of Norway Gro Harlem Brundtland, in 1987 (Kates, Parris, and Leiserowitz 2005). In that report ("the Brundtland Report"), sustainable development is defined in the following way (World Commission on Environment and Development 1987, 43):

> Sustainable development is development that meets the needs of the present without compromising the ability of future generations to meet their own needs. It contains within it two key concepts:
> 1. The concept of "needs", in particular the essential needs of the world's poor, to which overriding priority should be given, and
> 2. The idea of limitations imposed by the state of technology and social organisation on the environment's ability to meet present and future needs.

According to one researcher, a root cause of the public denunciation of this definition as vague, imprecise and not-operational is precisely the variety of interest groups battling for hegemony over the definition of the term "sustainable development" (Spangenberg 2002, 1). It is worth quoting Spangenberg's comments on the contested nature of the definition, if only to flag it up as an example of meaning finitism at work, especially the role of interests in defining the meaning of a term:

> This definition has been publicly denounced as vague, imprecise and not operational. However, it has helped to put sustainable development on the international policy agenda, making the concept a vision so impossible to ignore that all kinds of interest groups are actively battling for hegemony regarding the definition. No surprise then, that the approaches suggested lack any common definition or operational principle: a common definition of sustainable development needs to cover a broad range of interests which have no easily identifiable common denominator.
>
> So it is not the vagueness of the WCED's suggestion, but the conflicts of interest which cause the somewhat murky picture of what sustainability is all about. (2002, 1)

Elsewhere, Spangenberg remarks how the concept of sustainability is differently perceived in different European countries, based on their different cultural and political histories and traditions. Hence, Spangenberg (2007, 5-6) claims that in many parts of Central and Eastern Europe, among political circles the concept of sustainability was linked to social justice in relation with environmental resources; in Switzerland and Scandinavia, the term was linked to international justice; in France the sustainability debate centred on freedom, whilst in Germany and the Netherlands debates focussed on the environment. Once, again, it is noticed that the term is used differently by different linguistic, political or scientific communities, thus exemplifying meaning finitism at work.

According to Spangenberg, the value of sustainability can also be laid down in terms of three "imperatives" (2007):

- The environmental imperative: safeguarding the environment
- The social imperative: realising justice between people, countries, gender, social groups, etc.
- The institutional imperative: securing political participation

Accordingly, the "sustainability triple p challenge" of pollution, poverty and participation deficits can be identified as an aim for which science and technology are called upon to help with, even though it is recognised that they have played a role in the proliferation of these problems in the first place. Hence sustainability science has been developed from the outset as a science with clear non-epistemic goals, which is called upon to provide policy solutions to specific societal problems, drawing heavily on the resources of both the natural and the social sciences, as well as the humanities and philosophy. I will now move on to the relationship between sustainability science and policy.

6.3 Sustainability science and policy

6.3.1 Some considerations

In the previous section I elaborated on the political origins of the value of sustainability and in the challenges that sustainability science is called upon to provide solutions for. In this section I will elucidate more the relationship between sustainability science and policy, thus continuing the exploration of a theme that I addressed in Section 1.3 of the current work.

As remarked above, sustainability science is explicitly policy-oriented, and is often driven by political considerations and conducted by political think tanks[3]. Some further observations can be made on the subject-matter of sustainability science, as well as more broadly on environmental science. The first observation is that the model describing knowledge and policy production on matters regarding the environment follows Schema Two rather than Schema One as described in Chapter Four, in the sense that the process of mitosis has yet to take place, the "sides" of the controversy are not well-defined, and furthermore the need for a decision is pressing.

I further remarked that

> This is becoming increasingly pervasive as policy and state decision-making is more and more informed by and reliant on science. In such cases where uncertainty reigns supreme, it is easier for factors extraneous to science to influence the knowledge produced. (Chapter Four, Section 4.3.2).

I will now return to and elaborate on the workings and processes of policy-relevant science, flagging up sustainability science as a prime example.

As Jäger (1998, 143) remarks, citing a point made by Funtowicz and Ravetz (1991),

> issues of global environmental change differ from "traditional scientific problems", because they are global in scale and have long-term impacts, data are generally inadequate and the phenomena are complex and not well

---

[3] As evidence of the politics involved in sustainability science, Bettencourt and Kaur (2011) remark that the most dominant sources of publications in the field are Washington DC and London, something which is significant given that they are mostly known as political or administrative centres and not as leading university cities (although London can boast about some of the finest universities in the world)

> understood. Decisions have to be made on the basis of uncertain inputs and under somewhat urgent conditions.

In short, the "natural science" involved in environmental science and sustainability science projects, is more often than not "science-in-the-making" rather than settled science[4], and furthermore often the timeframe for the taking of a policy decision is quite short, with the pressures from outside groups and political and social interests being massive and sometimes overpowering the desire or the capacity of scientists to produce answers they are happy with.

Jäger (1998, 143) continues, citing another damning limitation for environmental policy-making:

> Funtowicz and Ravetz pointed out that the limitations of traditional problem-solving strategies in dealing with global environmental risks arise because decisions depend on evaluations of future states of the natural environment, resources and human society, all of which are unknown and unknowable.

Funtowicz and Ravetz continue to issue quite a bleak assessment5 of the efforts made to deal with uncertainty; they note that there have so far been two main strategies when dealing with uncertainty, situated at two extremes of a spectrum. The first approach has been the theoretical treatment, which has resulted in shrewd philosophical analyses of the relationship between knowledge and ignorance, as well as the risks involved. Such analyses, according to the authors, may achieve a high level of understanding, however they rarely yield practical tools suited for the quality evaluation of uncertain information. On the other hand, technical analyses of uncertainty and simple quality taxonomies have resulted in classifications of uncertainty specific to each discipline, with mathematical formalisms treating them as just another physical variable to be incorporated in computer models. The authors remark that the latter are ignored in practice even by the scientists dealing with the models themselves6.

---

[4] See Sections 2.5.1 and especially 4.3 for discussions on the distinction between settled science and science-in-the-making.

[5] As opposed to Shackley and Wynne (1996) who argue that talk of uncertainty in certain contexts serves to reinforce the authority of science along with a particular idiom of policy

[6] A similar criticism may be considered in the case of assessment methods that attempt to incorporate value considerations (for example Multi Criteria Assessment) into mathematical models.

Two main sources of uncertainty plaguing the input of the natural sciences into environmental and consequently sustainability science projects may be discerned from above. The first is the uncertainty deriving from the limited and inadequate character of the data collected, as related to the complexity and consequent lack of theoretical understanding of the systems studied. In other words, "wet" science concerning real-life problems and phenomena is much messier and ill-understood than "dry" science conducted in sterilised and well-controlled laboratory settings.7

The second source of uncertainty concerns more the policy aspect of decisions to be taken, and is more damaging as it is impossible to limit, as it concerns future states of the natural and social world. What we can do at best is to predict a future state of the system studied, however in this case, as opposed to quantum mechanical systems8 we are attempting to predict the future state of a system whose current state is ill-defined, ill-understood and very crudely modelled. This sort of uncertainty also has another damaging effect to the science-policy interface. It destroys the linear model of science policy in which first science is done, and then policy is produced based on the scientific output and evaluations. Rather than this, there are a lot more feedback loops between science and policy. Hence a more complicated model of the interaction between science and policy is sought for, an issue which I will address in the following section.

6.3.2 Some models for the science-policy interface

In Chapter One, more specifically in Section 1.3.2 I posed the question of whether science can solve societal problems, using Weinberg's (1972) distinction between scientific and trans-scientific questions. Furthermore, in laying down a model of egalitarian democracy which I advocate, in Section 1.5.3 I laid down a typology of the roles that the scientist may play in the advocacy for policy; I concluded with a suggestion for the correspondence of the roles of the lay-person, the expert adviser and the politician to the role of the chooser of aims, the choosers of possible means of achieving those aims, and the seeker of compromise in case of conflicts of opinion. I now recognise that this neat separation of roles may be too idealistic or in any case not be applicable to cases of real policy-

[7] A problem for traditional conceptions of science is that they too often focussed on the latter, using physics as the paradigmatic "dry" science. However, this is beginning to change as biology, which is often 'wet', is treated more and more as a paradigmatic natural science.

[8] And even in such system, only the simplest cases of systems of atoms are studied, as larger systems become mathematically intractable.

making as regards sustainability science projects, given that the problems addressed in such projects demand integrated solutions, and where the policy proposal is presented to political or societal bodies as one co-produced by a team consisting of both scientists, social and natural, as well as political and social interest groups and lay-stakeholders.

Having proposed communitarian epistemology and especially the view of the possessor of knowledge as a celebratory social status granted by the relevant community, I can now come back to a more sophisticated study of the relationship between science and policy. This is especially so since, in the case of sustainability science at least, the failure of the "linear" or "consequential" model that assumes that "first the science is done and then the policy is formulated and implemented" (Jäger 1998, 143) is evident.

6.3.2.1 Closure revisited

As regards the fine-grained model of closure, the communitarian epistemology-based explanation of decision-making based on science that I gave in Chapter Four implicitly assumed a neat separation between the scientific knowledge-claims grounding or justifying a policy decision and the decision itself. It also focussed only on how the knowledge-claim is produced and assimilated as knowledge by different communities, an assumption justified at the time as I was interested primarily in the production and dissemination of scientific knowledge, and only derivatively at decision-making based on science.

Taking a closer look at the decision-making process reveals that the neat separation between science and policy cannot be usefully maintained, at least in what concerns complex and uncertain subjects such as policy regarding climate change. The blending of science-based policy and scientific knowledge creation and dissemination, and the concept of regulatory science that is derived from it, sets off from a criticism of Weinberg's notion of "trans-science". The criticism, as articulated by Jasanoff (1987, 202-203) is that Weinberg sought to staunchly defend the Mertonian scientific ethos from attacks by "at least some of the most extreme sociologists of knowledge" (Weinberg 1986, 67), by claiming that it is the scientist's prerogative to decide where science ends and trans-science begins. According to Jasanoff, Weinberg advocates that issues of cognitive uncertainty are not intrinsic to science but rather emerge in the regulatory process, hence scientists themselves are allowed to preserve their professional authority and institutional power. Furthermore, Weinberg defends science from accusations of bias, subjectivity and disharmony, something which he views as rife in the policy-making process. However, he does claim that much of the "science" of probabilistic risk analysis is within the domain of trans-science, and furthermore explicitly shows the way towards "regulatory science" by claiming that one should "define a new branch of science, called regulatory science, in which the norms of scientific proof are less demanding than are the norms in ordinary science" (Jasanoff, 1987, 203).

Taking their cue from Weinberg, Shackley and Wynne (1995) develop the concept of "regulatory science", in which, according to Jäger (1998, 144)

> the content and context of research was subject to political pressures to provide answers, often in a short period of time, to a specific policy- or legally-driven question.

Regulatory science consequently does not involve new research, but rather consists of tools for evaluation and assessment. Instead of neatly separating science and policy, regulatory science is a hybrid where science and policy become inextricably linked.

An interesting feature that the two authors incorporate into the model of regulatory science, which is linked to the breakdown of the linear model is a process labelled "mutual construction" of science and policy, which is the notion that the domain of science

> helps to reinforce the belief that particular knowledge, ideas or "needs" in the policy field are realistic and valid, driven by policy-relevance and/or by the criteria defining "best science", and vice versa for the effects of policy on science (Shackley and Wynne 1995, 221) .

According to Jäger (1998, 144) the two authors also argue that closure on particular policy questions with a scientific element may often be achieved through pragmatic regulatory policy considerations, even though the crucial decisions may subsequently be presented as being only scientific in character. Shackley and Wynne, claims Jäger, also argue that climate change is a strong case of mutual construction in which the scientists' commitment to general circulation models and the expressed policy "needs" are mutually reinforcing.

Having laid down this explanation of the linkage between policy and science at a fine-grained level I will now go back to the proposal that sustainability science presents an exemplar of communitarian science. I will present three challenges to it, in the hope that I will make its content clearer by answering them.

6.4 Three challenges to sustainability science as communitarian science

6.4.1 Interdisciplinarity and communication

As elaborated in section 6.2.1, there is a lot of debate on how exactly to define the value that is sustainability. This is perhaps reflected on the fact that a lot of projects with an avowed interest in the value of sustainability would still consider themselves as interdisciplinary, rather than falling within the domain of the discipline of sustainability science. Indeed, Spangenberg et al. (2012) distinguish between the often disciplinary science for sustainability, which is considered more akin to traditional "applied science" and often results in products rather than solutions, and the interdisciplinary science of sustainability, or transformative science, which is closer to the model of communitarian democratised science/policy hybrid proposed above. The latter, according to Spangenberg et al. (2012) requires a wholesale transformation of the whole institutional fabric where scientific knowledge is created. In the remainder I will focus on a challenge directed at projects of such transformative science.

Vucetich and Nelson (2010, 539) give the following framework for the understanding of sustainability within academia:

> a. Development of efficient technologies and markets for meeting human needs, which is generally the purview of engineering, physical science, biotechnology, economics, and business;
> b. Understanding the state and nature of ecosystems, which is generally the purview of ecology and environmental science;
> c. Understanding how exploitation affects ecosystems, which is generally the purview of applied ecology and environmental science;
> d. Understanding how exploitation affects human cultures, which is generally the purview of sociology, political science, policy, law, anthropology, and the arts and humanities;
> e. Understanding the meaning of normative concepts such as human needs, socially just, depriving, and ecosystem health, which is generally the purview of ethics and philosophy.

To this they add a putative sixth category, that of understanding human needs, which they associate with human medicine and psychology. Given that in such projects there are also participants and stakeholders from outside academia, sustainability science projects display what three commentators label as "wide" and "deep" interdisciplinarity (Frodeman,

Mitcham, and Sacks 2001, 6), that is, such projects attempt to bridge both the natural science/humanities and the scientists/non-disciplinary public divides.

Hence it should not be of any surprise that the challenge that I will study is that of effective communication within the research team on the one hand, and on the other hand, communication between the research team and the different stakeholders of a given project.

I will briefly mention two possibilities that may ensure effective communication. The first is what Spangenberg (2011, 278) has labelled "disciplinary multilingualism", a concept very close to what Collins and Evans (2007), label as "interactional expertise". Collins describes interactional expertise "as the ability to converse expertly about a practical skill or expertise, but without being able to practice it, learned through linguistic socialisation among the practitioners" (2004, 125). Indeed, Collins includes interdisciplinary research projects as instances where interactional expertise is present. A criticism of the notion of interactional expertise related to interdisciplinary projects is that it is more suited to "narrow" interdisciplinarity, that is, it may be quite difficult and time-consuming to achieve such an expertise in projects where "wide" interdisciplinarity is required. Hence, it may be easier for an atmospheric physicist to develop interactional expertise in oceanography, or even for a sociologist (such as Collins, whose own research project involved spending time with the scientists he chose to study) to develop interactional expertise in gravitational wave physics than for a historian to develop interactional expertise in astrophysics. Collins, by demanding that the interactional expert be indistinguishable from the contributory expert save the fact that they are not practitioners, seems to set the bar quite high up, or in any case higher than what can be achieved within a project involving many disciplines and with a relatively short time frame – furthermore it seems to be also unsuited to accommodate cases where deep interdisciplinarity is involved. It would just be unrealistic to demand from a lay participant in a science of sustainability project to become interactional expert in the multitude of disciplines involved in order to truly co-produce knowledge and policy.9

The second possibility that I will briefly address is the forging of a single community and common language which may be unique to the project. "Creating" such a language may be easier than it seems, in the sense that it may be sufficient for the group members to hold some

[9] And perhaps even harder for the scientific expert to become interactional expert in certain systems of non-institutionalised and non-systematised knowledge.

meetings where they honestly lay down their interests10 in the project asides from the commitment to sustainability, responsibility and quality and through consensus reach agreement on what will constitute the aims of the project as well as their "lifeworld", also being local to the project.

By this I want to designate the following interpretation that Beyer (2011) gives of one of Husserl's key terms:

> If we consider a single community of subjects, their common lifeworld, or "homeworld", can be looked upon, by first approximation, as the system of senses or meanings constituting their common language, or "form of life" (Wittgenstein), given that they conceive of the world and themselves in the categories provided by this language.

Beyer also gives the following alternative definition of the lifeworld:

> If we consider subjects belonging to different communities, we can look upon their common lifeworld as the general framework, or "a priori structure", of senses or meanings that allows for the mutual translation of their respective languages (with their different associated "homeworlds") into one another. (2011)

The difference between the two definitions is to be found in whether one chooses, as given with the "interactional expertise" possibility of common language, to think of the members of the research team as a motley crew belonging primarily to the community of their disciplinary background, or, as per the "common language" possibility, as forming a community centred on the project and primarily be associated with that community.

It seems to me that the difference between the two definitions is also a difference between transformative science on the one hand, and good "normal" interdisciplinary science on the other, coupled at best with a pluralist version of the framing model as regards the science/policy interface. It is not impossible that such a mutual translation may take place, however, especially in the case of engaging with various different stakeholders such a process of translation would become so convoluted to the point where it would be incredible to assume that such a process is going on in the heads of all participants and stakeholders. I suspect that what may be happening in science of sustainability projects or in successful projects with deep and wide interdisciplinarity is that things may begin from mutual translation but progress to the agreement on a

[10] As these interests, if they stay hidden, may prevent the forming of a single community focussed on the project at hand, given lack of trust.

common finitist system of meanings local to the project. The contestation of concepts will still take place all the time; however as time passes and the group "gels" together, it will develop its own array of common exemplars which will ensure that the effects of this phenomenon will be minimal. Such a "language" would enhance the sense of belonging in a community and enhance the sense of co-ownership of any subsequent knowledge and proposal obtained.

6.4.2 Truth, reality and action

The question of the roles of truth and reality is bound to be present when one promotes a relativist epistemology, as I have done, especially given that I am proposing communitarian epistemology as regards the natural sciences, which are considered to be all about discovering truths about a mind-independent reality. In the case of sustainability science however, an answer is given the moment different disciplines and different perspectival viewpoints come together: a pluralist framework suggests itself, with all participants agreeing as to the existence of an external reality, whilst also claiming the following two assumptions: that methodologically, each discipline offers a different way of uncovering some desired truths about the external reality (perhaps even local to the project at hand), and that all methodologies are equal in this. This may be good as a starting assumption, and the relativist may be forced to accept the metaphysical assumption of there being a single external reality. However, I argue that the first of the two assumptions is untenable. My argument, which runs against the patchwork pluralist picture preferred by authors such as Dupré (1993, Cartwright (1999) and French and Costa (2003, chap. 5, 6) is that pluralism is an inherently unstable position, as it cannot deal adequately with the very real issue of contradictions between both scientific theories and utterances and examples from everyday life. Given that such contradictions arise, with the most famous being between the theory of relativity and quantum mechanics, the pluralist has to give up either one of the components of their position: either the single reality assumption – hence their position collapses into that of the relativist – or the assumption that there are many ways or equal perspectives of interacting with reality – hence their position collapses into standard realism. In a sense, pluralism has been exposed as wanting to both have the cake and eat it at the same time. As for my own position (which I have also laid down in Chapter Three of the present work), which starts from disagreement in the political and social sphere and wants to ground itself in democracy, the question between relativism and realism has to be framed as one between messy democracy and open-endedness on the one hand and elitism along with a privileged access to "truth" on the other; hence my preference for the relativist perspective.

However, if the relativism that I support needs any mitigation, then this is to be found in the constant negotiation and renegotiation in a democratic manner of concept meanings in a meaning finitist fashion and the setting aside of points of agreements as components of this language by granting them variously the statuses of "truth", "fact" and "knowledge". It is important to stress here that according to the full-blown meaning finitist, words such as "truth" and "fact" are themselves susceptible to a meaning finitist analysis (Kusch 2002, chap. 16) and as such they contain their own drifting array of exemplars and are perpetually contestable.

A further feature of communitarian science as presented here is its action-oriented character rather than its truth or even significant-truth (to use Kitcher's terminology) orientation. Truth or what the communitarian science project epistemic community regards as the true state of affairs is one of the determinants of the future state of affairs that is the aim of social action, along with the values and interests of the epistemic community. The important issue here for the success of a communitarian science project is the achievement of a strong commitment to participatory democracy as the mode of forging the common lifeworld of the project and the division of cognitive labour, as well as the mechanism by which knowledge is co-produced. In this sense, it is paramount that the threat of elitism is weeded out from the outset, a topic which I will now return to.

6.4.3 Sustainability science and elitism

As stated above, the challenge of a communitarian science project worth its salt lies in the following question: how will sustainability science avoid the formation of policy-making scientific elites, just like the sciences that I have so far been criticising? A gesture towards answering this question has been given through the description above of transformative science, however here I will attempt to elaborate a bit on how sustainability science as proposed can overcome this pitfall.

The distinguishing feature that sustainability science as communitarian science possesses are the participation of lay stakeholders both as members of the project team as well as those who are finally called upon to implement the final policy proposals; the co-authorship of the knowledge produced by the epistemic community that is formed by the project participants; and finally the organisation of the team on a participatory democratic fashion, that is in a fashion where all participants participate equally in knowledge creation.

I will briefly expand on the last point, that of all participants being equal. The main bone of contention is, as discussed in Chapter One, whether equality can co-exist with the division of cognitive labour. That there will be such a division of labour is unavoidable, given the complexity of the task. However, an answer to this has already been given in the above section on interdisciplinarity and communication. A common language particular to the project at hand ensures that the knowledge produced is, at the highest level, the product of a project "assembly" in which negotiations take place. In this assembly, the experts are "supervised" by the lay-people who have the final say in the implementation of the project recommendations (Feyerabend 1978, 96–98).

6.5 A plea for transformative science/policy

As a final word, I would like to once again stress on the one hand the need for a communitarian epistemology-inspired transformative science/policy hybrid, and on the other the opportunity that such a hybrid of participatory science/policy can offer to humanity. The description of sustainability science as transformative/communitarian science I have given above is meant to be a model for how all science may develop, both disciplinary and interdisciplinary. The key here is to make sure that scientific knowledge is regarded as a common good, produced by the citizens for the citizens. This has been my line all along. In order for this to happen, there needs to be a wholesale transformation in at least three aspects of science and scientific knowledge production; first, there needs to be a restructuring of school science education, in which, as argued for in Chapter Three, science is taught as a way of public argumentation. Furthermore, scientific findings, even at the highest level, should be freely available in a format that is understandable to lay-people. Secondly, at university level, there should be a subordination of (public) research money to public scrutiny, as well as a tighter link between university[11] knowledge production and societal problems. Finally, at the level of society at large, an effort should be made by lay-people to engage with science and to learn about the social elements involved in scientific knowledge production, as well as into the messy and uncertain character of participatory democracy.

[11] In order to avoid the scientization of politics, as well as the politicisation of scientific knowledge, what I am proposing here concerns all university production, not just that of the natural sciences.

The need for a transformative science is even more pronounced given the inequality in the access and use of natural resources. Such inequality, as well as widespread abuse of the natural environment, which often results in quasi-irreversible detrimental changes in our natural environment, is becoming increasingly a source of social, economic and political instability. Furthermore, we have reached a point where the existing unsustainable economic models have failed spectacularly to predict a world-wide economic crisis, and where unsustainable science produced through marketable product-oriented research has resulted in an irresponsible abuse of natural resources and the disappearance of pools of traditional sustainable knowledge. A wholesale different approach needs to be adopted, and transformative science seems to be an ideal candidate.

On the other hand, there is evidence of the beginnings of a move towards a science of sustainability, as well as the creation of institutions and large-scale funding schemes which result in the award of funds to projects which could be labelled as transformative science projects. More encouraging is the fact that society seems to be ready and demanding, often quite forcefully, a more participatory framework for the production and dissemination of knowledge as well as a transparent way of producing policy. Transformative science again seems like a very good candidate for satisfying both needs, through the co-production of knowledge and the collective shouldering of responsibility for policies produced.

6.6 Concluding remarks

6.6.1 What has been achieved

In the current work I attempted to bring together some disparate areas of enquiry such as political philosophy and work on democracy, research on the public understanding of science, and epistemology, to name a few. The main threads linking all these themes is the thesis that scientific knowledge is so important that it should be considered as a common good, and as such be available to all democratic citizens, and furthermore the processes of its production should be regarded as an exemplar of democracy at work. I have reasoned that the only epistemology that does justice to the notion of science as a permanent open-ended process of argumentation is communitarian epistemology, its relativistic implications notwithstanding. I embraced relativism together with a commitment to radical democracy, along with a focus on communities as the epistemic subjects, instead of individuals. The focus on communities has allowed me to enhance the democratic credentials of scientific knowledge production for the purposes of policy, by talking about the co-authorship and co-ownership of knowledge produced. I also embraced and defended meaning finitism, along with some of its most hard-to-stomach implications such as the finitist analysis of "truth" and "fact". Finally, in this chapter I presented, through the example of sustainability science and the notion of transformative science, an example of communitarian knowledge production for social and environmental policy.

6.6.2 Further related research

This brings me to the question of how such work can be continued. As already mentioned, one idea would be to give policy proposals that would help achieve my proposal for the adoption of communitarian epistemology for the enhancement of citizenship awareness among future citizens. An idea that I have considered is that of furthering the cycle of mandatory studies in order to allow future citizens to gain a basis in science and technology studies, together with a final examination in front of a lay audience, which will grant them citizen status as well as full voting rights.

Another domain which could do with some furthering would be science education and the social aspects of cognition, as described through theories such as activity theory. Even though I was considering engaging actively with such literature, at the end considerations of space dictated otherwise.

Finally, a more ambitious project linked to this project concerns standardisation and strengthening the defences of meaning finitism. Standardisation and the attempts to ground standards on an objective basis transcend the narrow field of metrology and could have widespread consequences on a range of philosophical questions. One such example of a study would be the study of standards of good vision and the historical relationship and influence of the printing press on what counts as "normal vision". Further examples would be the historical origins of designations of origin, or the questioning of the assumption that subatomic particles are identical with other particles of their class, rather than similar. All of these are projects that seem germane to my own research interests at this time.

REFERENCES

Armstrong, David Malet. 1973. *Belief, Truth and Knowledge*. London: Cambridge University Press. http://books.google.com/books?hl=en&lr=&id=NT04AAAAIAAJ&oi=fnd&pg=PA3&dq=belief,+truth+and+knowledge&ots=U15EEXg8d9&sig=mBNFc5w5wFE5SuDHSwZEB2QxCcE.

Audi, Robert. 2006. 'Testimony, Credulity, and Veracity'. In *The Epistemology of Testimony*, 25. oxford: Oxford University Press.

Baber, Zaheer. 1996. *The Science of Empire: Scientific Knowledge, Civilization, and Colonial Rule in India*. Albany: State University of New York Press.

Bacon, Francis. 1966. *New Atlantis*. World's Classics. Oxford: Oxford University Press.

Baker, G. P., and P. M. S. Hacker. 1984. 'On Misunderstanding Wittgenstein: Kripke's Private Language Argument'. *Synthese* 58 (3) (March): 407–450. doi:10.1007/BF00485249.

Barker, A., and B. G Peters. 1993. 'Introduction: Science Policy and Government'. In *The Politics of Expert Advice: Creating, Using and Manipulating Scientific Knowledge for Public Policy*, edited by Anthony Barker and B. Guy Peters, 1–16. Edinburgh: Edinburgh University Press.

Barnes, Barry. 1974. *Scientific Knowledge and Sociological Theory*. Routledge & K. Paul.

—. 1982. *T.S. Kuhn and Social Science*. Macmillan.

Barnes, Barry, David Bloor, and J. Henry. 1996. *Scientific Knowledge: A Sociological Analysis*. University of Chicago Press.

Barnett, Hilaire. 2004. *Constitutional & Administrative Law*. Routledge.

Basalla, George. 1988. *The Evolution of Technology*. Cambridge History of Science Series. Cambridge: Cambridge University Press.

Bauer, M.W., N. Allum, and S. Miller. 2007. 'What Can We Learn from 25 Years of PUS Survey Research? Liberating and Expanding the Agenda'. *Public Understanding of Science* 16 (1): 79–95.

Bettencourt, L.M.A, and J. Kaur. 2011. 'The Evolution and Structure of Sustainability Science'. *PNAS* 108 (49): 19540–19545.

Beyer, Christian. 2011. 'Edmund Husserl'. In *The Stanford Encyclopedia of Philosophy*, edited by Edward N. Zalta, Winter 2011.

http://plato.stanford.edu/archives/win2011/entries/husserl/.
Biletzki, Anat, and Anat Matar. 2011. 'Ludwig Wittgenstein'. In *The Stanford Encyclopedia of Philosophy*, edited by Edward N. Zalta, Summer 2011.
http://plato.stanford.edu/archives/sum2011/entries/wittgenstein/.
Billig, M. 1987. *Arguing and Thinking: A Rhetorical Approach to Social Psychology*. Cambridge: Cambridge Univ Pr.
Binmore, K. G. 2008. *Game Theory: A Very Short Introduction*. Oxford University Press.
Bird, Alexander. 2010. 'Social Knowing: The Social Sense of "Scientific Knowledge"'. *Philosophical Perspectives* 24 (1) (December): 23–56.
Bird, Alexander, and Emma Tobin. 2010. 'Natural Kinds'. In *The Stanford Encyclopedia of Philosophy*, edited by Edward N. Zalta, Summer 2010. http://plato.stanford.edu/archives/sum2010/entries/natural-kinds/.
Blackburn, S. 1984. 'The Individual Strikes Back'. *Synthese* 58 (3): 281–301.
Bloor, David. 1983. *Wittgenstein: A Social Theory of Knowledge*. Macmillan London.
—. 1996. 'Review Symposia'. *Metascience* 5 (1): 7–85.
—. 1999a. 'Anti-latour'. *Studies in History and Philosophy of Science* 30: 81–112.
—. 1999b. 'Reply to Bruno Latour'. *Studies in History and Philosophy of Science* 30: 131–138.
—. 2002. *Wittgenstein, Rules and Institutions*. London: Routledge.
Bohman, James, and William Rehg. 2011. 'Jürgen Habermas'. In *The Stanford Encyclopedia of Philosophy*, edited by Edward N. Zalta, Winter 2011.
http://plato.stanford.edu/archives/win2011/entries/habermas/.
Brakel, J. 1986. 'The Chemistry of Substances and the Philosophy of Mass Terms'. *Synthese* 69 (3): 291–324.
Brandom, Robert. 1994. *Making It Explicit: Reasoning, Representing, and Discursive Commitment*. Cambridge Mass.: Harvard University Press.
Bush, Vannevar. 1945. *Science, the Endless Frontier*. U.S. Govt. print. off.
Campbell, Richard. 1992. *Truth and Historicity*. Oxford: Clarendon press.
Cardwell, Donald. 1972. *Technology,science and History : a Short Study of the Major Developments in the History of Western Mechanical Technology and Their Relationships with Science and Other Forms of Knowledge.* London: Heinemann Educational.
Carr, Wilfred, and Anthony Hartnett. 1996. *Education and the Struggle for Democracy*. Buckingham and Philadelphia: Open University Press.
Cartwright, Nancy. 1983. *How the Laws of Physics Lie*. Oxford: Oxford

University Press.
—. 1999. *The Dappled World: a Study of the Boundaries of Science*. Cambridge UK ;New York NY: Cambridge University Press.
Chang, Hasok. 2004. *Inventing Temperature : Measurement and Scientific Progress*. Oxford ;New York: Oxford University Press.
Childs, John. 1989. 'The Educational Philosophy of John Dewey'. In *The Philosophy of John Dewey*, Third, I:419–443. The Library of Living Philosophers. La Salle, Illinois: Open Court Publishing Company.
Christiano, Thomas. 1996. *The Rule of the Many: Fundamental Issues in Democratic Theory*. Westview Press.
Christiano, Tom. 2008. 'Democracy'. In *The Stanford Encyclopedia of Philosophy*, edited by Edward N. Zalta, Fall 2008. http://plato.stanford.edu/archives/fall2008/entries/democracy/.
Churchland, Patricia. 1992. *A Neurocomputational Perspective: The Nature of Mind and the Structure of Science*. The MIT Press.
Churchland, Paul. 1989. *Neurophilosophy: Toward a Unified Science of the Mind-brain*. the MIT press.
Clark, Norman. 1985. *The Political Economy of Science and Technology*. Oxford: Basil Blackwell.
Clark, William C. 2007. 'Sustainability Science: A Room of Its Own'. *Proceedings of the National Academy of Sciences* 104 (6) (February 6): 1737 –1738. doi:10.1073/pnas.0611291104.
Cohen, Joshua. 1989. 'Deliberation and Democratic Legitimacy'. In *The Good Polity*, 67–92. Oxford: Blackwell.
—. 2003. 'Deliberation and Democratic Legitimacy'. In *Debates in Contemporary Political Philosophy- An Anthology*, 342–360. London and New York: Routledge.
Collins, Harry. 1985. *Changing Order: Replication and Induction in Scientific Practice*. London: Sage Publications.
—. 1997. 'Expertise: Between the Scylla of Certainty and the New Age Charybdis'. *Accountability in Research* 5 (1): 127–135.
—. 2004. 'Interactional Expertise as a Third Kind of Knowledge'. *Phenomenology and the Cognitive Sciences* 3 (2): 125–143.
Collins, Harry, and Robert Evans. 2007. *Rethinking Expertise*. University of Chicago Press.
Contessa, Gabriele. 2010. 'Scientific Models and Fictional Objects'. *Synthese* 172 (2).
Converse,, P.E. 1964. ' The Nature of Belief Systems in Mass Publics,'. In *Ideology and Discontent*, edited by D.E. Apter, 206–61.
Coombs, Rod, Paolo Saviotti, and Vivien Walsh. 1987. *Economics and Technological Change*. New Jersey: Rowman & Littlefield.

Cooper, R. 2004. 'What Is Wrong with the DSM?' *History of Psychiatry* 15 (1): 5–25.

Cooper, R.V. 2007. *Psychiatry and Philosophy of Science*. Acumen.

Costa, Newton C. A. da, and Steven French. 2003. *Science and Partial Truth: a Unitary Approach to Models and Scientific Reasoning*. Oxford University Press US.

Couch, Arthur, and Kenneth Keniston. 1960. 'Yeasayers and Naysayers: Agreeing Response Set as a Personality Variable.' *The Journal of Abnormal and Social Psychology* 60 (2): 151.

Craig, Edward. 1990. *Knowledge and the State of Nature: An Essay in Conceptual Synthesis*. Oxford: Clarendon Press.

Csaba, G., F. Birzele, and R. Zimmer. 2009. 'Systematic Comparison of SCOP and CATH: a New Gold Standard for Protein Structure Analysis'. *BMC Structural Biology* 9 (1): 23.

Danish Board of Technology. 2011a. 'Teknologirådet'. http://www.tekno.dk/subpage.php3?language=uk&page=forside.php3.

—. 2011b. 'The Vulnerability of the IT Infrastructure'. http://www.tekno.dk/subpage.php3?article=1091&language=uk&category=11&toppic=kategori11.

—. 2011c. 'Citizens Hearing'. http://www.tekno.dk/subpage.php3?article=816&toppic=kategori12&language=uk.

—. 2011d. 'The Consensus Conference'. http://www.tekno.dk/subpage.php3?article=468&toppic=kategori12&language=uk.

Davidson, Donald. 1991. 'Epistemology Externalized'. *Dialectica* 45 (2-3): 191–202. doi:10.1111/j.1746-8361.1991.tb00986.x.

Dawkins, Richard. 1996. *The Blind Watchmaker: Why the Evidence of Evolution Reveals a Universe Without Design*. 3rd ed. New York: Norton.

Day, R., D.A.C. Beck, R.S. Armen, and V. Daggett. 2003. 'A Consensus View of Fold Space: Combining SCOP, CATH, and the Dali Domain Dictionary'. *Protein Science* 12 (10): 2150–2160.

Dewey, John. 1981. 'My Pedagogic Creed'. In *The Philosophy of John Dewey*. Chicago: University of Chicago Press.

—. 2004. *Democracy and Education*. Dover Publications. http://books.google.com/books?hl=en&lr=&id=19ajcXf4MCYC&oi=fnd&pg=PA1&dq=democracy+and+education&ots=lGytcvNlF9&sig=5FrPvY_adNx0npMaedgs82iE_SE.

Diels, Hermann. 1903. *Die Fragmente Der Vorsokratiker. Griechisch Und Deutsch von Hermann Diels*. Weidmann.

Dohrenwend, Bruce P. 1966. 'Social Status and Psychological Disorder: An Issue of Substance and an Issue of Method'. *American Sociological Review*: 14–34.

Downs, A. 1957. 'An Economic Theory of Political Action in a Democracy'. *The Journal of Political Economy* 65 (2): 135–150.

Dretske, F. 1991. 'Two Conceptions of Knowledge: Rational Vs. Reliable Belief'. *Grazer Philosophiche Studien* 4: 15–30.

Dryzek, J. S., and C. List. 2003. 'Social Choice Theory and Deliberative Democracy: a Reconciliation'. *British Journal of Political Science* 33 (1): 1–28.

Dupré, John. 1993. *The Disorder of Things : Metaphysical Foundations of the Disunity of Science*. Cambridge Mass.: Harvard University Press.

Durant, J. 1993. 'What Is Scientific Literacy?' In *Science and Culture in Europe*, 129–138. London: Science Museum.

Erduran, Sibel, and Richard A. Duschl. 2004. 'Interdisciplinary Characterizations of Models and the Nature of Chemical Knowledge in the Classroom'. *Studies in Science Education* 40 (1): 105–138. doi:10.1080/03057260408560204.

Erduran, S., S. Simon, and J. Osborne. 2004. 'TAPping into Argumentation: Developments in the Application of Toulmin's Argument Pattern for Studying Science Discourse'. *Science Education* 88 (6): 915–933.

EU. 2002. 'Communication from the Communities on the Collection and Use of Expertise by the Commission: Principle and Guidelines (COM 2002:713),'. (COM 2002:713).

Eurobarometer. 2005. 'Social Values, Science and Technology'. European Commission.

Feyerabend, Paul. 1975. 'How to Defend Society Against Science'. *Radical Philosophy*. http://pagespro-orange.fr/tansen/bioethics/science/feyerabend.htm.

—. 1978. *Science in a Free Society*. London: NLB London.

—. 1993. *Against Method*. Third Edition. London: Verso.

—. 1999. 'Democracy, Elitism and Scientific Method'. In *Knowledge, Science and Relativism*. Cambridge: Cambridge University Press.

Fielding, Michael. 2008. 'On the Necessity of Radical State Education: Democracy and the Common School'. *Journal of Philosophy of Education* 41 (4): 539–557.

Fleck, Ludwik. 1981. *Genesis and Development of a Scientific Fact*. Translated by Thaddeus J. Trenn. University of Chicago Press.

Fodor, Jerry. 1974. 'Special Sciences, or the Disunity of Science as a Working Hypothesis'. *Synthese* 28: 77–115.

—. 1983. *The Modularity of Mind*. Cambridge, Mass: MIT Press.

—. 2001. *The Mind Doesn't Work That Way: The Scope and Limits of Computational Psychology*. The MIT Press.

—. 2008. *LOT 2: The Language of Thought Revisited*. Oxford University Press, USA.

Forge, J. 1996. 'Review Symposia'. *Metascience* 5 (1): 7–85.

Frigg, Roman. 2010a. 'Models and Fiction'. *Synthese* 172 (2): 251–268.

—. 2010b. 'Fiction and Scientific Representation'. In *Beyond Mimesis and Convention*, 97–138. http://link.springer.com/chapter/10.1007/978-90-481-3851-7_6.

Frodeman, R., C. Mitcham, and A.B. Sacks. 2001. 'Questioning Interdisciplinarity'. *Science, Technology, and Society Newsletter* 126: 1–5.

Fuller, Steve. 1997. *Science*. Buckingham: Open University Press.

Funtowicz, S. 2006. 'Why Knowledge Assessment?' In *Interfaces Between Science and Society*, edited by A.G. Pereira, S.G. Vaz, and S. Tognetti, 138–145. Sheffield: Greenleaf publishing.

Funtowicz, S., and JR Ravetz. 1991. 'Ecological Economics: The Science and Management of Sustainability'. In *Ecological Economics: The Science and Management of Sustainability*, edited by R Constanza. Columbia University Press.

Funtowicz, S., and R. Strand. 2007. 'Models of Science and Policy'. In *Biosafety First: Holistic Approaches to Risk and Uncertainty in Genetic Engineering and Genetically Modified Organisms*.

Galison, Peter Louis. 1997. *Image and Logic*. University of Chicago Press.

Garner, Bryan A. 2011. *Garner's Dictionary of Legal Usage*. Oxford University Press.

Gerrans, P. 1998. 'How to Be a Conformist, Part II. Simulation and Rule Following'. *Australasian Journal of Philosophy* 76 (4): 566–586.

Giere, R.N. 1990. *Explaining Science: A Cognitive Approach*. University of Chicago Press.

Gilbert, J.K., and C.J. Boulter. 1997. 'Learning Science Through Models and Modelling'. In *The International Handbook of Science Education*, 1:53. Dordrecht: Kluwer.

Gilbert, Margaret. 1992. *On Social Facts*. Princeton, New Jersey: Princeton Univ Press.

—. 2000. *Sociality and Responsibility: New Essays in Plural Subject Theory*. Lanham, Maryland: Rowman & Littlefield Pub Inc.

Gillett, Grant. 1995. 'Humpty Dumpty and the Night of the Triffids: Individualism and Rule-following'. *Synthese* 105 (2) (November): 191–206. doi:10.1007/BF01064218.

Goldman, Alvin I. 1967. 'A Causal Theory of Knowing'. *The Journal of*

Philosophy 64 (12): 357–372.
—. 1999. *Knowledge in a Social World*. Oxford University Press.
Greene, Brian. 2004. *The Fabric of the Cosmos: Space, Time and the Texture of Reality*. London: Penguin Books Ltd.
Gregory, Jane, and Steve Miller. 2000. *Science in Public: Communication, Culture, and Credibility*. Perseus Books Group.
Grosslight, L., C. Unger, E. Jay, and C.L. Smith. 1991. 'Understanding Models and Their Use in Science: Conceptions of Middle and High School Students and Experts'. *Journal of Research in Science Teaching* 28 (9): 799–822.
Habermas, Jurgen. 1971. 'The Scientization of Politics and Public Opinion'. In *Toward a Rational Society*, 62–80. London: Heinemann Educational Books.
Habermas, Jürgen. 1971. 'Reflections on the Linguistic Foundations of Sociology': The Christian Gauss Lectures (Princeton University, February–March 1971)'. *Published in Habermas, J.(2001) On the Pragmatics of Social Interaction*: 1–103.
Hacking, I. 1995. *Rewriting the Soul: Multiple Personality and the Sciences of Memory*. Princeton Univ Pr.
Hacking, Ian. 1999. *The Social Construction of What?* Cambridge, Massachussets: Harvard University Press.
Hadley, C., and D.T. Jones. 1999. 'A Systematic Comparison of Protein Structure Classifications: SCOP, CATH and FSSP'. *Structure* 7 (9): 1099–1112.
Hardwig, John. 1991. 'The Role of Trust in Knowledge'. *The Journal of Philosophy* 88 (12) (December): 693–708.
Harper, Simon. 2011. 'Conference Report from the EU Cities and Science Communication Conference – Science Policy Recommendations and How They Can Be Achieved | CASC'. http://blog.eucasc.eu/2011/02/conference-report-from-the-eu-cities-and-science-communication-conference-science-policy-recommendations-and-how-they-can-be-achieved/.
Harrison, A.G., and D.F. Treagust. 2000. 'A Typology of School Science Models'. *International Journal of Science Education* 22 (9): 1011–1026.
Hawking, Stephen, and Leonard Mlodinow. 2010. *A Briefer History of Time*. Transworld.
Hellman, G., and J.L. Bell. 2006. 'Pluralism and the Foundations of Mathematics'. *Editorial Board Ronald N. Giere (Philosophy, University of Minnesota) Keith Gunderson (Philosophy, University of Minnesota) Geoffrey Hellman (Philosophy, University of Minnesota)*: 64.

Hempel, C.G. 1965. 'Fundamentals of Taxonomy'. In *Aspects of Scientific Explanation and Other Essays in the Philosophy of Science*, edited by C.G. Hempel, pp. 137–154. Glencoe, Illinois: Free Press.

Hendry, R.F. 2006. 'Elements, Compounds, and Other Chemical Kinds'. *Philosophy of Science* 73 (5): 864–875.

Hesse, Mary. 1966. *Models and Analogies in Science*. Notre Dame, Indiana: University of Notre Dame Press.

Hesse, M.B. 1974. *The Structure of Scientific Inference*. Univ of California Pr.

Illich, Ivan. 1971. *Deschooling Society*. London: Calder & Boyars.

Ipsos MORI. 2011a. 'Public Attitudes to Science 2011'. London.

—. 2011b. 'Public Attitudes to Science 2011 - Literature Review'. London.

Jäger, J. 1998. 'Current Thinking on Using Scientific Findings in Environmental Policy Making'. *Environmental Modeling and Assessment* 3 (3): 143–153.

Jasanoff, S.S. 1987. 'Contested Boundaries in Policy-relevant Science'. *Social Studies of Science* 17 (2): 195–230.

Jenkins, Edgar. 1999. 'School Science, Citizenship and the Public Understanding of Science'. *International Journal of Science Education* 21 (7): 703–710.

—. 2007. 'School Science: a Questionable Construct?' *Journal of Curriculum Studies* 39 (3): 265–282.

Jones, Steve. 2003. *Y: The Descent of Men*. New York: Houghton Mifflin Harcourt.

Kates, Robert W., ed. 2010. *Readings in Sustainability Science and Technology. CID Working Paper No.213*. Cambridge, MA: Center for International Development, Harvard University.

Kates, Robert W., Thomas M. Parris, and Anthony A. Leiserowitz. 2005. 'What Is Sustainable Development? Goals, Indicators, Values, and Practice'. *Environment: Science and Policy for Sustainable Development* 47 (3) (April): 8–21.

Kealey, Terence. 1996. *The Economic Laws of Scientific Research*. Basingstoke and London: Macmillan Press Ltd.

Kemp, René, and Pim Martens. 2007. 'SSPP: Sustainable Development: How to Manage Something That Is Subjective and Never Can Be Achieved?' *Sustainability: Science, Practice & Policy* 3 (2). http://sspp.proquest.com/static_content/vol3iss2/0703-007.kemp-print.html.

Kemp, S. 2003. 'Toward a Monistic Theory of Science'. *Philosophy of the Social Sciences* 33 (3): 311–338.

Kitcher, Philip. 2001. *Science, Truth, and Democracy*. Oxford ;New York: Oxford University Press.

Knight, J., and J. Johnson. 1994. 'Aggregation and Deliberation: On the Possibility of Democratic Legitimacy'. *Political Theory* 22 (2): 277–296.

Knorr-Cetina, K. 1995. 'How Superorganisms Change: Consensus Formation and the Social Ontology of High-energy Physics Experiments'. *Social Studies of Science* 25 (1): 119–147.

Krosnick, Jon A., Allyson L. Holbrook, Laura Lowe, and Penny S. Visser. 2006. 'The Origins and Consequences of Democratic Citizens' Policy Agendas: A Study of Popular Concern About Global Warming'. *Climatic Change* 77 (1-2): 7–43.

Kuhn, D. 1992. 'Thinking as Argument'. *Harvard Educational Review* 62 (2): 155–179.

Kuhn, Thomas. 1962. *The Structure of Scientific Revolutions*. 3rd ed. Chicago: University of Chicago Press.

Kuhn, Thomas S. 1977. 'The Essential Tension. Selected Studies in Scientific Tradition and Change'. *Philosophy of Science, Chicago: University of Chicago Press,| C1977* 1. http://adsabs.harvard.edu/abs/1977etss.book.....K.

Kusch, Martin. 2002. *Knowledge By Agreement*. Oxford: Oxford University Press.

Lackey, J. 2005. 'Memory as a Generative Epistemic Source'. *Philosophy and Phenomenological Research* 70 (3): 636–658.

Laclau, E., and C. Mouffe. 1985. *Hegemony and Socialist Strategy: Towards a Radical Democratic Politics*. London: Verso. http://books.google.com/books?hl=en&lr=&id=-ZVoVtwCMz0C&oi=fnd&pg=PR7&dq=hegemony+and+socialist+strategy&ots=Q0STkQ-MVW&sig=p-SM5-rWzi1BTQPcaeB954kFVZU.

Latour, Bruno. 1987. *Science in Action: How to Follow Scientists and Engineers through Society*. Cambridge, Massachussets: Harvard University Press.

—. 1999. 'For David Bloor... and Beyond: A Reply to David Bloor's' Anti-Latour"'. *Studies in History and Philosophy of Science* 30: 113–130.

Laudan, Larry. 1978. *Progress and Its Problems: Towards a Theory of Scientific Growth*. Berkeley: University of California Press.

—. 1981. 'A Confutation of Convergent Realism'. *Philosophy of Science* 48 (1) (March): 19–49.

—. 1983. *Mind and Medicine: Problems of Explanation and Evaluation in Psychiatry and the Biomedical Sciences*. Vol. 8. Univ of California Pr.

Lau, L.Y., and R. Ranyard. 2005. 'Chinese and English Probabilistic

Thinking and Risk Taking in Gambling'. *Journal of Cross-Cultural Psychology* 36 (5): 621–627.

Lee, S. 1996. 'Cultures in Psychiatric Nosology: The CCMD-2-R and International Classification of Mental Disorders'. *Culture, Medicine and Psychiatry* 20 (4): 421–472.

Lehrer, Keith. 1987. 'Personal and Social Knowledge'. *Synthese* 73 (1): 87–107.

Lenoir, Timothy. 1997. *Instituting Science*. Stanford, California: Stanford University Press.

Liberatore, Angela. 1993. 'Chernobyl Comes to Italy - The Reciprocal Relationships of Radiation Experts, Government Policies and the Mass Media'. In *The Politics of Expert Advice: Creating, Using and Manipulating Scientific Knowledge for Public Policy*, edited by Anthony Barker and B. Guy Peters, 33–47. Public Administration in the 1990s. Edinburgh: Edinburgh University Press.

Linus Publications. 2009. 'Linus Publications'. http://www.linusbooks.com/.

Locke, J. 1975. *An Essay Concerning Human Understanding (1690), Ed.* Edited by PH Nidditch. Oxford: Clarendon Press.

Logan, R. A. 1992. 'Popularization Versus Secularization: Media Coverage of Health'. In *Risky Business: Communicating Issues of Science, Risk, and Public Policy*, 43–59.

Luhmann, N. 1992. 'What Is Communication?' *Communication Theory* 2 (3): 251–259.

MacFarlane, J. 2007. 'Relativism and Disagreement'. *Philosophical Studies* 132 (1): 17–31.

Macgill, Sally. 1993. 'Controversy and Authority in British Official Scientific Advice on Radiation - The Black Report on Sellafield and Children's Leukaemias'. In *The Politics of Expert Advice: Creating, Using and Manipulating Scientific Knowledge for Public Policy*, edited by Anthony Barker and B. Guy Peters, 61–90. Public Administration in the 1990s. Edinburgh: Edinburgh University Press.

Margolis, Eric, and Stephen Laurence. 2011. 'Concepts'. In *The Stanford Encyclopedia of Philosophy*, edited by Edward N. Zalta, Fall 2011. http://plato.stanford.edu/archives/fall2011/entries/concepts/.

Mayes, R., and A.V. Horwitz. 2005. 'DSM-III and the Revolution in the Classification of Mental Illness'. *Journal of the History of the Behavioral Sciences* 41 (3): 249–267.

McDowell, J. 1984. 'Wittgenstein on Following a Rule'. *Synthese* 58 (3): 325–363.

McGinn, Colin. 1984. *Wittgenstein on Meaning: An Interpretation and*

Evaluation. Oxford; New York: Blackwell.

Mermin, N.D. 1998. 'The Science of Science'. *Social Studies of Science* 28 (4): 603–623.

Miller, Fred. 2012. 'Aristotleś Political Theory'. In *The Stanford Encyclopedia of Philosophy*, edited by Edward N. Zalta, Fall 2012. http://plato.stanford.edu/archives/fall2012/entries/aristotle-politics/.

Mill, John Stuart. 1993. 'Representative Government'. In *Utilitarianism: On Liberty ; Considerations on Representative Government ; Remarks on Bentham's Philosophy*. London: Dent.

Mokyr, Joel. 2002. *The Gifts of Athena*. Princeton, New Jersey: Princeton University Press.

Mouffe, Chantal. 2000. 'For an Agonistic Model of Democracy'. In *The Democratic Paradox*, 143. London: Verso.

National Science Board. 2012. *Science and Engineering Indicators 2012*. Arlington VA: National Science Foundation.

Needham, P. 2000. 'What Is Water?' *Analysis* 60 (1): 13–21.

Nelkin, Dorothy. 1977. *Technological Decisions and Democracy*. Beverly Hills: Sage Publications.

—. 1984. *Science as Intellectual Property: Who Controls Research?* Macmillan.

Nie, Norman H., Jane Junn, and Kenneth Stehlik-Barry. 1996. *Education and Democratic Citizenship in America*. University of Chicago Press.

Nowotny, Helga, Peter Scott, and Michael Gibbons. 2001. *Re-thinking Science: Knowledge and the Public in an Age of Uncertainty*. Wiley-Blackwell.

Nunes, Terezinha, Analúcia Dias Schliemann, and David William Carraher. 1993. *Street Mathematics and School Mathematics*. Cambridge University Press.

O'Connor, C. 1993. *Traditional Cheesemaking Manual*. ILRI (aka ILCA and ILRAD).

Oppenheim, Paul, and Hilary Putnam. 1958. 'Unity of Science as a Working Hypothesis'. In *Minnesota Studies in the Philosophy of Science- Concepts, Theories, and the Mind-Body Problem*. Vol. II. Minneapolis: University of Minnesota Press.

Osborne, J., S. Erduran, and S. Simon. 2004. 'Enhancing the Quality of Argumentation in School Science'. *Journal of Research in Science Teaching* 41 (10): 994–1020.

Parsons, Talcott. 1965. 'Max Weber 1864-1964'. *American Sociological Review* 30 (2) (April): 171–175.

Paulhus, Delroy L. 1991. 'Measurement and Control of Response Bias.' In *Measures of Personality and Social Psychological Attitudes*, 17–59.

San Diego, CA: Academic Press, Inc. http://doi.apa.org/psycinfo/1991-97206-001.

Peabody, Dean. 1966. 'Authoritarianism Scales and Response Bias.' *Psychological Bulletin* 65 (1): 11.

People Science, and Policy Ltd/TNS. 2008a. 'Public Attitudes to Science'.

—. 2008b. 'Public Attitudes to Science - A Guide'.

Perry, W.G. 1970. 'Forms of Intellectual and Ethical Development'. *New York: Holt, Rinehart et Winston.*

Phillips, Derek L. 1973. *Abandoning Method*. San Francisco: Jossey-Bass. http://library.wur.nl/WebQuery/clc/367293.

Pickstone, John. 2000. *Ways of Knowing : a New History of Science, Technology, and Medicine*. Manchester: Manchester University Press.

Pielke Jr., Roger A. 2007. *The Honest Broker: Making Sense of Science in Policy and Politics*. Cambridge: Cambridge University Press.

Pittas Dairy Industries Ltd. 2012. 'Pittas Dairy Industries Ltd'. http://www.pittas.com/default.aspx?articleID=13932&heading=History.

Plato. 2004. *Plato - Gorgias*. Translated by Walter Hamilton and Chris Emlyn-Jones. Revised Edition. Penguin Classics. London: Penguin Books.

Psillos, S. 2011. 'Living with the Abstract: Realism and Models'. *Synthese* 180 (1): 3–17.

'Public Praises Science; Scientists Fault Public, Media'. 2009.

Putnam, Hilary. 1987. 'Truth and Convention: On Davidson's Refutation of Conceptual Relativism'. *Dialectica* 41 (1-2): 69–77.

—. 1990. *Realism with a Human Face*. Cambridge, Mass. ; London. Harvard University Press.

Qualitative Study. 2008. 'The Image of Science and and the Research Policy of the European Union'. European Commission. http://ec.europa.eu/research/science-society/document_library/pdf_06/qualitative-study-rapport08012009_en.pdf.

Quine, Willard Van Orman. 1960. *Word & Object*. MIT Press.

Ravetz, J. R. 2011. 'Postnormal Science and the Maturing of the Structural Contradictions of Modern European Science'. *Futures* 43 (2): 142–148.

Rawls, John. 2005. *Political Liberalism*. Columbia University Press.

Remm, M., C.E.V. Storm, and E.L.L. Sonnhammer. 2001. 'Automatic Clustering of Orthologs and In-paralogs from Pairwise Species Comparisons1'. *Journal of Molecular Biology* 314 (5): 1041–1052.

Republic of Cyprus. 2009. *Protected Designation of Origin Application*.

Restivo, Sal. 1988. 'Modern Science as a Social Problem'. *Social Problems* 35 (3) (June): 206–225. doi:10.2307/800619.

Richards, Keith, ed. 2010. *The New Optimists: Scientists View Tomorrow's World and What It Means to Us*. Birmingham: BPR Publishers.

Rolin, K. 2008. 'Science as Collective Knowledge'. *Cognitive Systems Research* 9 (1-2): 115–124.

Rosch, E., and C. B Mervis. 1975. 'Family Resemblances: Studies in the Internal Structure of Categories'. *Cognitive Psychology* 7 (4): 573–605.

Rosenhan, D.L. 1973. 'On Being Sane in Insane Places'. *Science* 179 (4070): 250.

Rounsaville, B.J., K. Bryant, T. Babor, H. Kranzler, and R. Kadden. 1993. 'Cross System Agreement for Substance Use Disorders: DSM-III-R, DSM-IV and ICD-10'. *Addiction* 88 (3): 337–348.

Ryder, Jim. 2002. 'Science, Citizens and Schools: Opportunities and Challenges'. *Studies in Science Education* 37 (1): 156–162. doi:10.1080/03057260208560183.

Saoulli, Alexia. 2007. 'Halloumi Bickering Threatens EU Application'. *The Cyprus Mail*, March 3. http://web.archive.org/web/20070813101718/http://www.cyprus-mail.com/news/main.php?id=31047&archive=1.

Sarewitz, Daniel. 2004. 'How Science Makes Environmental Controversies Worse'. *Environmental Science & Policy* 7 (5) (October): 385–403. doi:10.1016/j.envsci.2004.06.001.

Saward, M. 2000. 'Less Than Meets the Eye: Democratic Legitimacy and Deliberative Theory'. In *Democratic Innovation: Deliberation, Representation and Association. ECPR Studies in European Political Science.*, edited by M. Saward, 66–77. London and New York: Routledge. http://oro.open.ac.uk/16396/1/LessThanMeetsTheEye.pdf.

Schattschneider, E.E. 1960. *The Semisovereign People*. New York: Holt, Rinehart and Winston.

Schrader, D.E. 2004. 'Intellectual Safety, Moral Atmosphere, and Epistemology in College Classrooms'. *Journal of Adult Development* 11 (2): 87–101.

Schumpeter, Joseph Alois. 1976. *Capitalism, Socialism and Democracy*. London: George Allen & Unwin Ltd.

Searle, John R. 1995. *The Construction of Social Reality*. London: Penguin Books.

—. 2006. 'Social Ontology Some Basic Principles'. *Anthropological Theory* 6 (1): 12–29.

Shackley, Simon, and Brian Wynne. 1996. 'Representing Uncertainty in Global Climate Change Science and Policy: Boundary-ordering Devices and Authority'. *Science, Technology & Human Values* 21 (3):

275–302.

Shackley, S., and B. Wynne. 1995. 'Global Climate Change: The Mutual Construction of an Emergent Science-policy Domain.' *Science and Public Policy* 22 (4): 218–230.

Shapin, Steven. 1994. *A Social History of Truth: Civility and Science in Seventeenth-century England*. University of Chicago Press.

Silverman, David. 2006. *Interpreting Qualitative Data: Methods for Analyzing Talk, Text, and Interaction*. SAGE.

Sokal, Alan. 2008. *Beyond the Hoax: Science, Philosophy and Culture*. Oxford: Oxford University Press.

Spangenberg, Joachim. 2002. 'Sustainability Science: Which Science and Technology for Sustainable Development'. *INES Newsletter* 39: 9–14.

—. 2007. 'Towards An Integrated Concept of Sustainability'. In Goethe Institute, Riga.

—. 2011. 'Sustainability Science: a Review, an Analysis and Some Empirical Lessons'. *Environmental Conservation* 38 (03): 275–287.

Spangenberg, Joachim, Haris Shekeris, Georgina Cundill, and Joshua Cinner. 2012. 'En Route to Transformative Science: The Challenges of and to Interdisciplinary Sustainability Science'.

Special Eurobarometer. 2005. 'Europeans, Science and Technology'. 224. European Commission.

—. 2010. 'Science and Technology'. European Commission. http://ec.europa.eu/public_opinion/archives/ebs/ebs_340_en.pdf.

Spitzer, R.L. 1975. 'On Pseudoscience in Science, Logic in Remission, and Psychiatric Diagnosis: A Critique of Rosenhan's" On Being Sane in Insane Places".'

Spotlight Radio. 2012. 'Spotlight Radio'. Text. http://www.spotlightradio.net/listen/how-mad-are-you/.

Standard Eurobarometer. 2001. 'Europeans, Science and Technology'. European Commission.

Sturgis, P., and N. Allum. 2004. 'Science in Society: Re-evaluating the Deficit Model of Public Attitudes'. *Public Understanding of Science* 13 (1): 55.

Szasz, T.S. 1960. 'The Myth of Mental Illness.' *American Psychologist* 15 (2): 113.

Thagard, P. 2008. 'Mental Illness from the Perspective of Theoretical Neuroscience'. *Perspectives in Biology and Medicine* 51 (3): 335–352.

Thomas, Jeff. 2000. 'Issues for the Schooling of Science'. In *Science and the Citizen: For Educators and the Public*, edited by Roger T. Cross and Peter J. Fensham. Fitzroy, Victoria, Australia: Arena Publications.

Toulmin, S.E. 1958. *The Uses of Argument*. Cambridge Univ Pr.

Toumey, Chris, John Besley, Meg Blanchard, Mark Brown, Michael Cobb, Elaine Howard Echlund, Margaret Glass, Thomas M. Guterbock, A. Eammon Kelly, and Bruce Lewenstein. 2010. 'Science in the Service of Citizens & Consumers'. NSF.

Tournon, J. 1993. 'The State as Blind Investor in Fundamental Research – Advice to the French and Other Governments on the European Synchrotron Radiation Facility'. In *The Politics of Expert Advice: Creating, Using and Manipulating Scientific Knowledge for Public Policy*, edited by Anthony Barker and B. Guy Peters, 33–47. Public Administration in the 1990s. Edinburgh: Edinburgh University Press.

Tuomela, Raimo. 2003. 'Collective Acceptance, Social Institutions, and Social Reality'. *American Journal of Economics and Sociology* 62 (1): 123–165.

Tuomela, Raimo, and Wolfgang Balzer. 1998. 'Collective Acceptance and Collective Social Notions'. *Synthese* 117 (2) (January 1): 175–205.

Turnbull, D. 1996. 'Review Symposia'. *Metascience* 5 (1): 7–85.

UNEP. 1992. 'Rio Declaration on Environment and Development'. http://www.unep.org/Documents.Multilingual/Default.asp?documentid=78&articleid=1163.

U.S. Department of Commerce - Patent and Trademark Office, and Trademark Trial and Appeal Board. 1999. 'Danish Dairy Board, Inc. v The Ministry Of Commerce and Industry of the Republic of Cyprus'. Cancellation No. 19815. U.S. Government. http://www.uspto.gov/web/offices/com/sol/foia/ttab/other/1999/1999.htm.

Van Cleve, J. 2006. 'Reid on the Credit of Human Testimony'. In *The Epistemology of Testimony*. Oxford: Oxford University Press.

Verba, Sidney, Kay Lehman Schlozman, and Henry E. Brady. 1995. *Voice and Equality: Civic Voluntarism in American Politics*. Harvard University Press.

Votsis, Ioannis. 2012. 'Gastronomia'. http://www.votsis.org/gastronomia.html.

Vucetich, J.A., and M.P. Nelson. 2010. 'Sustainability: Virtuous or Vulgar?' *BioScience* 60 (7): 539–544.

Vygotskiĭ, Lev Semenovich. 1978. *Mind in Society: The Development of Higher Psychological Processes*. Translated by Michael Cole. Harvard University Press.

Wall, DP, HB Fraser, and AE Hirsh. 2003. 'Detecting Putative Orthologs'. *Bioinformatics* 19 (13): 1710–1711.

Walton, Douglas N. 1996. *Argumentation Schemes for Presumptive Reasoning*. Mahwah, NJ: Laurence Erlbaum Associates.

Walton, Kendall. 1990. *Mimesis as Make-believe: On the Foundations of the Representational Arts*. Harvard Univ Pr.

WCED World Commission on Environment and Development. 1987. *Our Common Future*. Oxford: Oxford University Press.

Weinberg, A.M. 1986. 'Science and Its Limits: The Regulator's Dilemma'. *Hazards: Technology and Fairness, National Academy Press, Washington DC*: 9–23.

Weinberg, A. M. 1972. 'Science and Trans-science'. *Minerva* 10 (2): 209–222.

Welbourne, Michael. 1986. *The Community of Knowledge*. Aberdeen University Press.

Wikipedia. 2012a. 'Protected Geographical Status'. *Wikipedia*. http://en.wikipedia.org/wiki/Protected_Geographical_Status.

—. 2012b. 'Halloumi'. *Wikipedia*. http://en.wikipedia.org/wiki/Halloumi.

—. 2012c. 'Bogdanov Affair - Wikipedia, the Free Encyclopedia'. *Bogdanov Affair*. January 13. http://en.wikipedia.org/wiki/Bogdanov_affair.

Wilce, Alexander. 2006. 'Quantum Logic and Probability Theory'. http://plato.stanford.edu/entries/qt-quantlog/.

Williamson, Timothy. 1995. 'Is Knowing a State of Mind?' *Mind* 104 (415): 533–565.

—. 2000. *Knowledge and Its Limits*. Oxford: Oxford University Press.

Wimsatt, W. 1976. 'Reductionism, Levels of Organization and the Mind-body Problem'. In *Consciousness and the Brain*, edited by Gordon Globus, Grover Maxwell, Irwin Savodnik, and Edmond M. Dewan. New York: Plenum Press.

Wittgenstein, L. 1953. 'Philosophical Investigations, Trans'. *GEM Anscombe* 261: 49.

Wolpert, Lewis. 1997. *The Unnatural Nature of Science*. Harvard University Press.

—. 2005. 'The Medawar Lecture 1998: Is Science Dangerous?' *Philosophical Transactions: Biological Sciences* 360 (1458) (June 29): 1253–1258.

Woodruff, E., and K. Meyer. 1997. 'Explanations from Intra-and Inter-group Discourse: Students Building Knowledge in the Science Classroom'. *Research in Science Education* 27 (1): 25–39.

World Health Organization. 1993. *The ICD-10 Classification of Mental and Behavioural Disorders - Diagnostic Criteria for Research*. Geneva.

—. 2012. 'History of the Development of the ICD'. www.who.int/classifications/icd/en/HistoryOfICD.pdf.

Wynne, B. 1992. 'Misunderstood Misunderstanding: Social Identities and

Public Uptake of Science'. *Public Understanding of Science* 1 (3) (July 1): 281–304. doi:10.1088/0963-6625/1/3/004.

Zachar, P. 2000. 'Psychiatric Disorders Are Not Natural Kinds'. *Philosophy, Psychiatry, & Psychology* 7 (3): 167–182.

Zheng, Y.P., K.M. Lin, J.P. Zhao, M.Y. Zhang, and D. Yong. 1994. 'Comparative Study of Diagnostic Systems: Chinese Classification of Mental Disorders-versus DSM-III-R'. *Comprehensive Psychiatry* 35 (6): 441–449.

Ziman, John. 1976. *The Force of Knowledge: The Scientific Dimension of Society*. Cambridge University Press.

—. 2000. *Real Science : What It Is, and What It Means*. Cambridge ;;New York: Cambridge University Press.